"Stretching"
Exercises for
Qualitative
Researchers
THIRD EDITION

I dedicate this text to all of my teachers, colleagues, friends, and mentors, who continue to inspire me. Likewise, this is dedicated to my students and all who helped me create this text in one way or another.

"Stretching" Exercises for Qualitative Researchers

THIRD EDITION

Valerie J. Janesick
University of South Florida

Los Angeles | London | New Delhi
Singapore | Washington DC

For information:

SAGE Publications, Inc.
2455 Teller Road
Thousand Oaks,
  California 91320
E-mail: order@sagepub.com

SAGE Publications India Pvt. Ltd.
B 1/I 1 Mohan Cooperative
  Industrial Area
Mathura Road, New Delhi 110 044
India

SAGE Publications Ltd.
1 Oliver's Yard
55 City Road
London EC1Y 1SP
United Kingdom

SAGE Publications Asia-Pacific Pte Ltd
33 Pekin Street #02-01
Far East Square
Singapore 048763

Printed in the United States of America

Library of Congress Cataloging-in-Publication Data

Janesick, Valerie J.
 "Stretching" exercises for qualitative researchers/Valerie J. Janesick. — 3rd ed.
   p. cm.
 Includes bibliographical references and index.
 ISBN 978-1-4129-8045-6 (pbk.)
  1. Social sciences—Research—Methodology. 2. Observation (Scientific method)
 3. Qualitative reasoning. I. Title.

H62.J346 2011
300.72—dc22   2010020251

This book is printed on acid-free paper.

10   11   12   13   14   10   9   8   7   6   5   4   3   2   1

Acquisitions Editor:	Vicki Knight
Associate Editor:	Lauren Habib
Editorial Assistant:	Ashley Dodd
Production Editor:	Catherine M. Chilton
Copy Editor:	Pam Schroeder
Typesetter:	C&M Digitals (P) Ltd.
Proofreader:	Annette R. Van Deusen
Indexer:	Molly Hall
Cover Designer:	Gail Buschman
Marketing Manager:	Stephanie Adams
Permissions:	Adele Hutchinson

Brief Contents ■

Contents ■

Preface ■

A great deal has occurred in the field of qualitative research since the second edition of this book in 2004, let alone since its inception in 1998. In fact, thinking about a third edition was a personal journey of stretching, writing, and reflection on experience. As I reflect on my own career in teaching qualitative research methods to doctoral students, this third edition is rewritten to walk through each additional evolutionary step in constructing our field of study as it appears today. Those of you who used the second edition know that I like to reimagine and create spaces for integrating the arts into the research process. In fact, I see the work of the qualitative researcher most like that of the artist. I use the metaphor of dance and stretching once again with incorporation of yoga as a metaphor in a new light. I have been studying yoga for nearly two decades now, and it is very close to dance in many ways. While the stretching metaphor is obvious in both, each relies on a strong history and at the same time a strong commitment to change, adaptation, and metamorphosis. This edition will have many new components in this metamorphosis. For example, like the choreographer who adapts using the methods of the day, in this new edition, I am adding a lengthy section on *Internet inquiry*. This is our epoch after all. It would be difficult to read a journal or book today without inclusion of the awareness of the power and possibility of the digital techniques, equipment, and presentation modes available to us. As researchers documenting the lived experience of participants, digital recorders and digital video camcorders for immediate uploading to YouTube, for example, can assist qualitative researchers in a way previously unknown. At the same time, the explosion of Internet sites, social networking groups, and blogs, while offering us new media to stretch, to observe, and to test interview strategies, also serve to raise questions of propriety and ethics in qualitative research.

■ Poetic Devices

In addition, I will concentrate more fully on *narrative and poetic writing, analysis,* and *presentation of data.* Also new in this book, I will rely on the theoretical frame of John Dewey's habits of mind. If we look at our work as observers, interviewers, reflective journal writers, photographers, and web-based connoisseurs, as habits of mind, the dancer's and philosopher's idea of "practice makes perfect" prevails. Likewise, I am integrating in this edition an overall exercise of writing a research reflective journal upon doing all the exercises and to develop that habit. At the end of using this text, my goal is for you to practice narrative writing in the researcher reflective journal. Consequently, this third edition, for me, is indeed stretching further. In another sense, this text is about *critical literacy* and *writing*. Observations must be understood through a critical lens while documenting and writing. Interview transcripts eventually are reinvented in new forms, such as poetry, as a critical edge in representation of data and pushing writing into a poetic mode. The researcher reflective journal is all about critical writing, thinking, synthesizing, and communicating. The use of photography as a research tool enriches our stories as well. Thus, the text is recast with an eye to history, to the digital era, to art and poetry, and to writing skills and development, with one foot still in the trenches of day-to-day negotiations with the Institutional Review Boards (IRBs) and various public stakeholders who are our partners. It is a working document, a work in progress, and always changing as my teaching is always changing.

I have been teaching qualitative research methods since my doctoral studies at Michigan State University. There, I first became engaged in classes on John Dewey's work with professors in the Philosophy Department and the College of Education. Dewey's work on art as experience, experience as education, and habits and will struck a chord with me that resonated with my work as a qualitative researcher and choreographer. His elegant thoughts on habits of mind have always inspired me. Since that time and throughout my career, I have been fortunate to work with gifted and articulate doctoral students who have raised questions, challenged my thinking along with their own beliefs, and have contributed to our knowledge base. It is exciting to be part of a corner of history in terms of shaping an agenda for and about our educational knowledge base, our theory construction, and our practice. I hope to continue to add to our field by incorporating many new stretching exercises to test and develop the observation and writing habit, the interviewing and writing habit, the creative habit, and the collaborative habit, which include keeping a researcher reflective journal, writing up our stories, writing up ethical dilemmas from the field, and finding new resources, such as the Internet, to assist us in our work. I have not abandoned the existing exercises, for they naturally set the stage for developing good habits of mind and researcher muscles as the researcher is the research instrument in true qualitative work. By using the arts, specifically poetry and photography, as components of qualitative research projects, I hope this third edition extends and strengthens our understanding of content knowledge, theory, and practice. In this text, I will expand my use of the metaphor of dance and yoga as a continuous lens from which to understand qualitative research methods.

■ Teaching Qualitative Methods

After completing my studies at Michigan State and earning my PhD, I immediately began teaching a trial course in qualitative methods to doctoral students in education and humanities with great interest and success in my first faculty role. After piloting the course, I took it through all bureaucratic channels to get it officially listed in the curriculum. In subsequent settings, even when a qualitative research course was on the books, it was most often taught by a quantitative researcher. You can imagine the discussions at those tables for both student and teacher. We lived through the 1980s and 1990s with these discussions. Indeed, I found then and continue to find today many doctoral students who have a default notion of research, which came to them through the many quantitative courses required of them. They automatically want to prove something or control something in the research setting. So, part of teaching qualitative research methods is working upstream against this default notion. I am totally in favor of students getting a balanced menu of courses to understand inquiry. Doctoral students need to know how to read both qualitative and quantitative research reports. They also need to know which questions are suited to qualitative research methods. Too often, I find that students actually want to do a survey and try to fit survey-type questions into an interview. Of course, it does not work. That is why I am writing this book. I want to expand how we think about research in general and qualitative research in particular. We need to focus on the questions suited to the appropriate method before selecting a means. Thus, in my own background in dance, drama, and the humanities, I was able to construct exercises for my students to sharpen their skills in the two major techniques of qualitative research: interviewing and observation. In this third edition, I added writing exercises to enrich the qualitative researcher in developing the habit of writing and the habit of using the mind for analysis and interpretation of the data.

In fact, every university that was my home was a setting for a repeat of this scenario. I went through every channel to propose not only one qualitative course but an advanced course as well. Most often, at least one of these would be accepted. At present, I am again in the process of trying to get a class on qualitative case methods through the various levels of bureaucracy, so I am once again in between technocratic channels. For me, this illustrates how watchful one has to be when attending to practical matters without losing sight of the purpose of doing research in the first place. I hope to widen our repertoire of classes beyond Qualitative Methods I and II. Currently, I am happy to say that, after more than 30 years of this struggle, today, in most doctoral programs at least, there are two or more qualitative research methods courses for students. In addition, interdisciplinary courses in other colleges are readily accepted. Thus, persistence pays off after all. In fact, persistence and patience are required of qualitative researchers before, during, and after the days in the field. Students find that this resonates with their goals and questions of interest. Students want and need qualitative research methods courses to refine their skills, as the researcher is the research instrument. Because qualitative researchers are research instruments, they want to practice and refine their skills in observation, interviewing, writing, analysis, and self-awareness. They will also be better served in the job market by being able to read qualitative research studies and discriminate between average and above-average studies, just as they are reputed to be able to do by taking five or more quantitative courses. I also think that students understand their use of statistics better by studying qualitative methods and vice versa. In fact, I have them read the great statistics teacher, Joel Best's work.

Best (2004) has done a service for qualitative researchers. He critiques his own field and statistical reasoning as well as the public fascination with numbers. His description of missing numbers, confusing numbers, scary numbers, authoritative numbers, magical numbers, and contentious numbers helps qualitative researchers in training. Students say that this book illuminates for them what numerous statistics classes have not. As a result, they understand better that, in their work as qualitative researchers, the words they use matter, and they will not use words to confuse, frighten, or omit relevant data so that a wider audience will get the meaning they intend. Furthermore, they want to be better at interviewing, observing, and writing. It is as simple and as clear as that. I continue to recreate and reinvent my classes. I change texts and practice exercises on a regular basis. It keeps all of us on our toes. At the same time, as I teach this every semester and as I continue to take yoga classes, I am changing as a person, teacher, scholar, and writer. How could I not change everything in my classes on a regular basis?

■ Progress in the Field

Since the first edition of this text, the context of our field has been illuminated by more than a dozen handbooks related to qualitative research methods and by new online and hard-copy journals numbering well over three dozen. These journals include and name many of the approaches of qualitative researchers: narrative research, arts-based research, action research, ethnographic research, oral history, life history, autoethnography, biography, autobiography, case study research in all its forms, phenomenological studies, grounded theory studies, feminist research approaches, netnography, and Internet inquiry to name the most prominent. In addition, websites, LISTSERVs, blogs, social network sites, and Internet dissertation coaches, for example, have widened the repertoire of making overt all the resources that assist qualitative researchers. Furthermore, the number of conferences devoted to qualitative research projects is only growing. Many key professional organizations have divisions and special interest groups devoted to qualitative research methodology. In fact, their major journals welcome qualitative work specifically. All in all, it is a good time to test the boundaries of and to develop new and better ways to do qualitative research. I hope this edition contributes to emphasizing stretching and development.

In this new edition, I have reconstructed some of the exercises from the second edition and added new exercises. New examples are used from current works throughout the text and in the appendixes. New in this edition, you will find the following:

1. A focus on developing good habits of mind, including the observation habit, the interview habit, the writing habit, keeping a researcher reflective journal habit, the creative habit, the analysis and interpretation of data habit, and the construction of poetry habit.

2. A focus on keeping a researcher reflective journal, and in fact, by the end of the book, the reader should have a substantial researcher in training portfolio and researcher reflective journal: its possibilities, its problems, the ethical implications of using digital settings in research, the use of the web, LISTSERVs, social networking sites, interviews online, software products, and exercises for practicing digital techniques.

3. Expanded information on the habit of doing observations and interviews in general, especially the need to pilot the interviews

and to use new forms of representation of interview data in the form of poetry, with new exercises for observation and interviews.

4. An integration of the habit of writing in multiple forms, such as narrative writing, poetry, and practice, with the researcher reflective journal as a data set.

5. A section on strategies for the qualitative researcher to interact with IRBs effectively.

6. Making sense of qualitative data through developing analytical habits that include creating categories and themes as well as models and a discussion of analysis of qualitative research data.

7. New samples and examples of student work, including the use of found data poems, that is, poetry created with words from transcripts and documents.

8. Electronic and other resources and how to contact them.

9. New exercises, including the oral history interview.

10. New additions to the appendixes to emphasize analysis, data category development, poetic devices, and a new consent form suitable to this decade's IRB requirements.

11. Exercises for completion of a researcher reflective journal by you, the reader, following from the exercises here.

In this version of the book, as in the previous editions, the arts function as the core metaphor. My life has evolved this way. From dance, which really has never left me, I have become a serious practitioner of yoga and meditation. In fact, I hope someday soon to become a teacher of yoga and meditation. It is a natural evolution for me, given my background in dance and qualitative research. You will see many references to the arts and to yoga here. There is integrity to movement in dance and in yoga that resonates with the entire shape of qualitative research techniques and processes. As one of my teachers told me, "Movement never lies: Once you understand your body you understand your mind." Similarly, the word *practice* is used to describe yoga and meditation as a practice. This resonates with qualitative research as a practice in itself. Qualitative research skills and techniques need development on a daily basis. This is like practicing yoga postures called *asanas*. Meditation is a reflective practice akin to the thinking time qualitative researchers need to make sense of the data collected.

I initially wrote this text for my current and future students. I originally had in mind doctoral students and faculty who would be qualitative researchers. Since the first printing, I have received numerous letters, e-mails, testimonials, and questions about these exercises from many individuals in other fields. One group that has regularly used this text, in addition to doctoral students in many fields, is students who will become teachers. It turns out that teacher-educators found this text valuable for practicing observing and interviewing, two techniques most useful to the would-be teacher. Moreover, they use these exercises for classroom research projects, sometimes called *action research* or *teacher research*. In addition, many have taken to calling the writing up of some of the case studies *critical case studies* (and I agree). It is a perfect fit with the observation and interview exercises in this text. Most of my letters and e-mails are from teachers who tell me how much this book has helped them with their research projects in the classroom. I have also heard from social workers, nurses, law students, and medical students. These exercises seem to be working for many outside of education and outside of the research community as well.

Again, this is very much like the arts, for one never knows, once a work of art is out there, who will attach themselves to it and find meaning in it. Nevertheless, this book has been written for any person interested in learning how to observe and see what is in front of them. It is for any person who wants to know how to interview, listen, and communicate with another human being. It is also the case that, because the qualitative researcher is the research instrument, one has to sharpen that instrument by training the eyes to see, the ears to hear, and the mouth and body to communicate. Like a dancer and yoga practitioner would do, one can achieve this only by practice, discipline, diligence, and perseverance. Reimagining this text has been, for me, a process of discovery. It took on a life of its own, and the result is before you. I hope you will find these exercises inspiring and useful as you move on your journey toward becoming a better qualitative researcher, a better educator, a better writer, and a better person.

Like the yoga teacher who strives for mindfulness and the dancer and choreographer who strive for unity in movement, this book is a testimonial. It is an autobiographic snapshot of my life work and its influences. For me, this book is a written record meant to inspire and prompt my students to lead a better life. After all, as the poet Henry Austin Dobson said, "All passes. Art alone enduring stays to us."

■ Acknowledgments

First and foremost, to my current and former students, I must say thanks for taking part in this adventure and always raising the bar by using imagination. To colleagues, friends, reviewers, and critics, I am most grateful. Those who contributed to the text with samples of their works in progress are generous, and I thank them for their graciousness. They are Patricia Williams-Boyd, Carolyn Stevenson, Kristy Loman Chiodu, Jason Pepe, Charles Bradley, Oksana Vrobel, Derya Kulavoz-Onal, Carol Burg, and Ruth Slotnick. Special thanks to my yoga and meditation teachers at the Deepak Chopra Center, Chicago Yoga Center, and Tampa Yoga Café for always inspiring me to stretch further, to think further, to observe further, and to see what is before me. After a yoga or meditation class, I always see the world in a new way. Finally, thanks to Vicki Knight, my editor, and all at Sage who made this process challenging and caring. I am grateful.

About the Author ■

Valerie J. Janesick (PhD, Michigan State University) is Professor of Educational Leadership and Policy Studies, University of South Florida, Tampa. She teaches classes in qualitative research methods, curriculum theory and inquiry, foundations of curriculum, issues in curriculum, ethics, and educational leadership. Her text, *"Stretching" Exercises for Qualitative Researchers* (2004), Sage, includes ways to integrate the arts in qualitative research projects. Now she introduces a third edition to add the technology connections and the use of poetry to represent interview data. Her writings have been published in *Curriculum Inquiry, Qualitative Inquiry, Anthropology and Education Quarterly*, and other major journals. Her chapters in the *Handbook of Qualitative Research* (first and second editions) use dance and the arts as metaphors for understanding research. She is completing oral history interviews of female school superintendents as part of a larger project on women leaders. Her recent chapter on Dewey and the arts and education can be found in the *Handbook of the Arts in Qualitative Inquiry: Perspectives, Methodologies, Examples, and Issues* (2007), Sage. Her latest book is *Oral History for the Qualitative Researcher: Choreographing the Story* (2010), Guilford Press.

She serves on the editorial boards of *Educational Researcher, The Journal of Curriculum and Pedagogy*, and *The Qualitative Report*. She is currently taking classes in yoga and meditation. Her most prized possession is her British Library Reader's Card, as she is working on an archival project on the letters of John Dewey written to educators around the world. Find her blog on critical pedagogy and poetry at: http://freire.mcgill.ca/content/critical-pedagogy-and-poetry

Correspondence to: vjanesic@usf.edu

Website (in progress): http://sites.google.com/site/valeriejjanesick/

CHAPTER 1

Qualitative Research and Habits of Mind ■

Story is far older than the art of science and psychology and will always be the elder in the equation no matter how much time passes.

—*Clarissa Pinkola Estes*
Women Who Run With the Wolves *(1996)*

Because the researcher is the research instrument in qualitative research projects, it is important for the researcher to practice and refine techniques and *habits of mind* for qualitative research. Habits of mind in this text will include observation habits, interview habits, writing the researcher reflective journal habit, the narrative writing habit, the habits of analysis and interpretation, and the habit of writing poetry found in the interview transcripts. In addition, the creative habit and the collaborative habit will also be discussed. Combining and working on all these habits will also prepare you to be a reflective researcher in terms of Internet inquiry along with the issues surrounding the use of blogs as data and representing data in visual format and visual text, all with an eye toward the ethical questions embedded in our research approaches. You need to fine-tune your observation skills, your interviewing skills, and your narrative and poetic writing skills, and this edition of the book will provide exercises to assist you in this journey on the path to being a better qualitative researcher. In addition, in this technology-centered world, you need to know how to use and critique the latest technological artifacts that may or may not make your work easier. In fact, you need to practice like a dancer warming up. You need to reflect on what works for you, like a yogi who meditates and stretches before class begins. All these bodily activities help to jump-start your brain, the primary center for you, the research instrument. Thus, you need to practice seeing what is in front of you. You need to practice hearing the data, that is, listening skills to hear what your participants are telling you. You need to flex your arm muscles and write in both narrative and poetic forms. In other words, you have to be present in the study. Like the dancer and practitioner of yoga, you need the body to reflect what the mind can produce. In this third edition, I rely on continuing work in the classroom and in the field along with my ongoing dialogue and debate with my students. As a teacher of qualitative research methods, I have been fortunate to have been both inspired and tested by my doctoral students. No matter what the geographical setting, the questions they have raised remain nearly the same:

1. How can I become a better qualitative researcher?

2. How can I improve observation skills?

3. How can I improve interview skills?

4. How can I become a better writer?

5. What can I do with these skills?

Most recently, another question has been forthcoming:

6. Can I get a job as a qualitative researcher? (Of course!)

Realizing that the first five questions have no easy answers and realizing that there is no one way to respond to these questions, I am using the concept of *stretching exercises* once again to frame this third edition as a response to my students' questions. I have written elsewhere (Janesick, 1994, 2000, 2001, 2007) using dance as a metaphor for qualitative research design and would like to extend that metaphor by using the concept of stretching. Stretching implies that you are moving from a static point to an active one. It means that you are going beyond the point at which you now stand. Just as the dancer must stretch to begin what eventually becomes the dance, the qualitative researcher may stretch by using these exercises to become better at observation and interview skills, which eventually solidify as the research project. These are meant as a starting point, not a slavish set of prescriptions. Also, in yoga, stretching is critical because you stretch not just the body but the mind as well. In fact, in yoga, every cell is activated by a series of *asanas*, or postures. The idea is that by activating your cells through stretching, breathing, and successive postures, you not only stretch the body but you also stretch emotionally, mentally, and spiritually. As a qualitative researcher in training, so to speak, you will grow in many ways as well. In fact, no matter how many tweets, twitter accounts, Facebook postings, blogs, or web-based activities you the researcher in training participate in, in the end, you still need to rely on yourself as the research instrument and rely on the two basic techniques of all qualitative work, observation and interviewing. In this book, you will find exercises to assist you in your development and hopefully assist in your definition of your role as a researcher.

■ Identifying Habits

As a starting point, I see these exercises as part of shaping the prospective researcher as a disciplined inquirer. Disciplined inquiry, a term borrowed from the renowned educator and philosopher John Dewey (1859–1952), assumes that we begin where we are now and, in a systematic way, proceed together to experience what it might mean to inquire. Furthermore, his notion of "habits of mind" resonates here. In this case, practice exercises are used to help in identifying a disciplined inquiry approach and to develop habits of mind. To use ballet as an example, the ballet dancer in training takes a series of beginning classes, intermediate classes, and various levels of advanced classes before going to performance *en pointe*. There is no way an individual can skip from beginning to advanced stages in ballet, or in modern dance, for that matter. In fact, the modern dance student is often required to study ballet at advanced levels in order to have a stronger ability to do modern dance. So it is with yoga. You can imagine that in a discipline where your goal is to integrate breath, body, mind, and spirit, you must proceed through levels of beginning, intermediate, and advanced study. It is the same for you, the qualitative researcher in training. You work toward developing as a qualitative researcher by developing habits of mind to enable you to move toward your goal.

One has to develop habits and skills and train the body incrementally. Likewise, the qualitative researcher has to train the mind, the eye, and the soul together as a habit. By doing these exercises, we allow for an interchange of ideas and practice, self-reflections, and overall evaluation of one's own progress through each of the chapters described in this text: (a) the observation and writing habit; (b) the interviewing and writing habit; (c) the creative habit; and (d) the analysis habit, which include making sense of the data, ethics, intuition, interpretation of data, Internet inquiry, and interaction with the Institutional Review Board (IRB). By assuming a posture of disciplined inquiry and assuming development of sturdy habits of mind, the prospective qualitative researcher is an active agent. This is not about memorizing a formula. Nor is it about dropping into a research project and finishing up quickly. This is about constructing a critical space for serious observation and interview skills habits and development of those habits. By actually constructing this space, the prospective qualitative researcher automatically begins a labor-intensive and challenging journey. This is like the journey of a dancer from first dance class to performance, or a student of yoga from beginning stretches to amazing handstands, backbends, headstands, and other postures that truly test the body and mind at every known level. This requires time, patience, diligence, ingenuity, and creativity, all of which are required for the qualitative researcher. Just as a dancer keeps track of movement phrases and critiques on performance, in the exercises described here, you will also keep track in a researcher reflective journal throughout the use of this book to reflect upon your own habits of mind as they develop.

■ Getting Feedback and Writing About It

Consequently, some exercises here may also provide a way to work on the role of the researcher by helping researchers to know themselves better. This can be helpful only when researchers are engaged full-time with participants in the field. Participants will trust the researcher if the researcher trusts himself or herself. It goes without saying that the researcher must have a solid knowledge of the self. One of the great things about graduate study is that if done correctly the student should grow and change remarkably. I have often remarked that graduate school is for stretching the mind and using parts of the brain that have not been used so actively previous to graduate study. One of the great things about teaching is watching that process unfold. Also, I like to think of the classroom as something of a studio space, a laboratory if you prefer, to see the various levels of habits of mind displayed in behavior. Graduate school or any learning space allows for the practice of the exercises and allows for space to fail, to receive feedback, and to redirect. Just as in a dance studio or yoga studio, constant feedback is given to allow for progression toward performance, so too in the studio of the qualitative research class-room, is there a space for critique, feedback, and redirection and not just from the instructor but from fellow researchers in training. The feedback loop is essential in this process. The dancer as artist is accustomed to the constant feedback loop of performance, critique and feedback, redirection, and then a new performance. After that performance, we do it all again with feedback and redirection to the next level. Another good example to illustrate this point can be seen in the performances at the Olympics. Judges assess the athletes through a rating system. They give feedback with that score, and then the coaches of the athletes go in-depth into the redirection for the next time around. The athlete has to take part in the redirection or suffer in the ratings. All of this conveys a sense of action, dynamism, and movement forward. So now, we can begin on the habits we need to develop to become a qualitative researcher and refine existing habits.

■ Finding Your Theoretical Habit

Research is an active verb. It is a way of seeing the world that goes beyond the ordinary. Thus, this series of practice exercises is designed to help learners along the way. In any class, there are beginners, midlevel learners, and advanced learners. The exercises written here help individuals find themselves in terms of their levels of expertise in observation, interviewing, writing, and analysis. This also requires self-knowledge.

These exercises are not a formula. I created them, and I have tested them for years. They resonate with students, and they work for me, provided students make the effort to really go the distance. In that sense, the qualitative researcher starts with two basic questions:

- What do I want to know?
- What set of techniques do I need to find out what I want to know?

This is the starting point for any research project. In qualitative work, the fact that the researcher is the research instrument requires that the senses be fine-tuned. Hence, the idea of practice on a daily basis sharpens the instrument. Many individuals can look at something and not see what is there. It is my goal to have readers of this text try to sharpen the following skills:

- Seeing through the habit of observation
- Hearing through the habit of interviewing

- Writing as a habit (researcher reflective journal, narrative writing, and poetry)
- Conceptualizing and synthesizing as a habit of mind (developing models of what occurred in the study)
- Communicating through ordinary language
- Reflecting on Internet resources and tools that may help in refining yourself as the research instrument

One of my dance teachers once said that dancers are, in the end, architects of movement. So, before an architect builds a building, he or she must understand, for example, the use of structure and the grammar of architecture, steel, plastic, and stone. Likewise, the qualitative researcher must understand the functions and feel of observations, interviews, writing, and so on, before the final written report of the study is created. Prior to that, the researcher also needs to know the theoretical foundations that guide his or her research. In this book, exercises are designed to allow for finding your theoretical framework. What is the *ology or ism* that guides you? Is it phenomenology? Is it social reconstructionism? Is it feminism? Is it critical theory in any of its forms? Your job by the end of this book and at the end of your researcher reflective journal artifact is to identify and describe the theory that guides your work and why and how you might use that to inform your observations, interviews, and writing.

I have found that learners respond to being actively involved in these practice exercises for a number of reasons. They have told me that these exercises strengthen their confidence, imagination, and ability to cope with field emergencies. (See Appendix B for reflective journal samples.) In addition, students appreciate the fact that all of these exercises are understandable, because the language used to describe them is ordinary language. I have always found students more responsive and enthusiastic when ordinary language is used to include them in the active engagement of qualitative research. They are more excited about theory, practice, and praxis when they are not excluded from the conversation. The reader of this text will see that the exercises are described in ordinary language, following in the tradition of bell hooks (1994), who pointed out that any theory that cannot be used in everyday conversation cannot be used to educate. In addition, the actual experience and practice of these exercises in observation and interview activities often help to allay fears and misconceptions about conducting qualitative research projects. Four of the most common misconceptions stated in classes or workshops are the following:

- Doing qualitative research is easy, and anyone can do it. Many students actually tell me that they are sent into my class because they were told it is easier than doing statistics. They then discover how qualitative research projects are time intensive and labor intensive.
- You should only learn qualitative research methods to augment quantitative work. In fact, this is a most woodenheaded notion. I find myself constantly working upstream to counter this default notion of research, that research must first be quantitative.
- Most people can do interviews and observations with little or no practice. Because in the workplace an individual may interview someone at the Auto Vehicles Bureau, there is an assumption that all interviewing is the same. By contrast, in the dance world, one would never attempt a performance without years of training, practice, and movement through beginning, intermediate, and advanced-level class work.
- Anyone can write a journal with no practice, preparation, or quiet time. Writing takes time and practice and should be a daily ritual if you are to effectively communicate your findings from a given study.

In a sense, this text is a response to these four misconceptions. I see these exercises as an opportunity to continue the conversation with individuals who want to take the plunge and develop strong habits of mind that allow for developing observation and interview skills. This of course means developing habits of mind for vigorous descriptive writing and powerful poetic writing if need be. This is one approach to learn qualitative research methods. Although I have used these exercises with doctoral students in education and human services, there is the continuing hope and possibility that beginning researchers at the master's degree level may also find these exercises useful, not only in education and human services but in other disciplines as well.

■ Developing Habits of Mind

I engage learners in these exercises over a period of 15 weeks, the artificial constraint of the given semester time line. In the best of all possible worlds, I would prefer a year or more, perhaps 3 semesters or even 3 courses of work time. I try to divide the course over the 15 weeks into 5 major habits:

1. *The Observation and Writing Habit.* Here, observation and narrative writing are connected, demanding time in terms of preparation and implementation and space for feedback and rewriting.

2. *The Interview and Writing Habit.* This includes both narrative and poetic writing. Interviews may be rendered in poetry constructed from the data, found data poems, or evocative poems, sometimes called evocative texts prompted by all the data sets.

3. *The Reflective Journal Writing Habit.* This allows for specifying all that encompasses your role as the research instrument.

4. *The Analysis and Interpretation Habit.* Here, time is needed for thinking, rewriting, and eventually constructing a model of what occurred in your study. This may also involve critical use of software or Internet data resources.

5. *The Creative Habit.* This is where intuition and creativity come into play and should be documented in addition to detailed documentation of any ethical issues that arise in a given study.

Learners simultaneously design, conceptualize, and conduct a ministudy. Within the study, the learner must conduct interviews and observations and keep a researcher reflective journal. During the first third of the class, learners practice exercises in and out of class that focus on observation, writing, getting feedback, rewriting, and the role of the researcher. The next third of the class is spent on interviewing exercises and practice in ongoing analysis. The learners also share their journals-in-progress with one another in small groups. Here, students may wish to create poetry from the interview data from documents received from participants in the study and from their own researcher reflective journals. The final third of the class is devoted to the final analysis of data, issues regarding intuition in research, and ethical issues. Ideally, the observation exercises and role of the researcher exercises are practiced for about 5 or 6 weeks, at which point the learners are asked to develop a plan in writing for their ministudies. Then, as they go into the field, they work simultaneously on interview exercises and role of the researcher exercises for the remainder of the semester, all the while reading selected books and articles in the area. They immediately begin to keep a written record in the form of a researcher reflective journal in order to keep track of their thoughts about the readings and engagement with the written text. This also prepares them for entry into the field. Although this series of exercises grew out of my work with students and workshop members, surely others may be interested in this book. Likewise, the reader of this text can be immersed in these exercises at any pace he or she may choose. However, as my meditation teacher once told me, "it takes thirty days to develop a habit." Similar to John Dewey's concept, the learner has to develop the habits of mind that allow for being better at writing, interviewing, and observing. Also, in the social world, because of the vast number of hours spent at the computer, online, video gaming, Facebooking, tweeting, and all the investment in Internet time frames, prospective researchers are using parts of the brain that do not typically engage what is needed to observe and *see* or to interview and *hear*. I am almost convinced that even creativity may be at risk if web-based activities distract from development of the research instrument, you, the researcher. Basically, this book is an attempt to begin a conversation with individuals who may not have had the time, energy, or interest in practicing qualitative techniques but are now ready to jump into the studio of the social world and stretch. Of course, there are always complications. For example, just the number of terms used to refer to qualitative research alone could confound anyone. Take a look at the list in the following section.

■ Terms Used to Describe Qualitative Research

The terms on the next page are just those I find myself using when working with students on their research projects. It is not exhaustive. The point of listing them is to alert the reader to the numerous terms. Just as there is no one way to learn ballet or no one way to learn yoga, there is no one way to learn how to be a qualitative researcher. Find your personal velocity. Which approach resonates with the person you are? When you know that, learn all you can about that approach. By now, there are many examples in textbooks, articles, the World Wide Web, and dissertation abstracts to name a few resources. Then, like the dancer and the yogi, practice, practice, practice. Practice also includes feedback, critique, redirection, and rewriting. For example, I took my first yoga class in 1976, a hatha yoga class. Since then, I have taken classes in Bikram yoga, Ashtanga yoga, Iyengar yoga, and kundalini yoga. Yet after all this, I have returned to totally immerse myself in hatha yoga. I returned to my starting point. So, from my experience, I suggest you find your niche from the approaches listed above. It is surely not meant to be all inclusive but rather a starting point for understanding the complexity of qualitative research approaches. I refer you to *Qualitative Inquiry: A Dictionary of Terms* by Thomas A. Schwandt (2001) for more definitive information. In addition, qualitative work has certain characteristics, such as the following.

Some Terms Used to Identify Qualitative Work

Action research	Historiography	Narrative research
Webography	Interpretive interactionist study	Oral history
Case study		Autoethnography
Descriptive study	Life history	Ethnoarcheology
Ecological study	Netnography	Teacher research
Narrative study	Interpretive policy analysis	Philosophical analysis study
Ethnography		
Field research	Microethnography	Portraiture

Furthermore, in qualitative work, certain questions resonate with the techniques and approaches we use. By the questions we ask, we are suited to qualitative work. In the following list of characteristics of qualitative work, you may see a reflection of the kind of questions suited to you, the qualitative researcher.

■ Characteristics of Qualitative Work

1. It is holistic: It attempts to understand the whole picture of the social context under study; in education at least, we have trampled on this unmercifully. Often individuals call their work qualitative when in fact it is not. This is due to the fact that they ask questions ill suited to any of the main qualitative approaches. They do not use the foundational frameworks for qualitative work either.

2. It looks at relationships within a system or subculture: Again, this relates to the holistic nature of qualitative work. There is nothing qualitative about doing a survey on SurveyMonkey and adding one question that requires a sentence or two. You will simply get a sentence or two of narrative, verbal, self-report data. Yet time and again, individuals will call their studies qualitative without meeting the rigorous standards of qualitative work.

3. It refers to the personal, face-to-face, immediate interactions in a given setting.

4. It is attentive to detail and focused on understanding the social setting rather than predicting and controlling.

5. It demands equal time in the field and in analysis; often individuals rush to meet a deadline and in their haste forget to analyze what is in front of them.

6. It incorporates a complete description of the role of the researcher; too often this is forgotten or recklessly done. The description of the role of the researcher must incorporate the biases, beliefs, and values of the researcher up-front in the study. In addition, the actual specifics should be described as to the number and types of observations and interviews, length of interviews, how transcripts were completed, how documents were collected and used, and so on.

7. It relies on the researcher as the research instrument.

8. It incorporates informed consent documentation and is responsive to ethical concerns in the study.

9. It acknowledges ethical issues in fieldwork with a complete discussion of these issues.

10. It considers, in many cases, participants as coresearchers in the project.

11. It tells a story in narrative or poetic forms.

12. It is useful for the reader of the research in terms of the coherence, cohesion, insight, and actual words of the participants.

I must return here to the metaphors of yoga and dance. Dance as an art form is one of the most rigorous and demanding of the arts. For one thing, physical tone and health are critical to the survival of the dancer. The hours of working out are not just physical, because the physical and mental connections engage the dance artist totally. The dancer's life is short in terms of performance due to dependence on a superbly functioning instrument— the body. The only practice that is more demanding is the serious study of yoga. I mention all of this to punctuate the fact that the discipline and desire of the dancer and the yoga student are persistent and indomitable, much like the qualitative researcher. As a professor of qualitative research methods, as a former choreographer and dancer, and now as a student of yoga, I see the role of the researcher as one characterized by discipline, persistence, diligence, creativity, and desire to communicate the findings so as to reflect the social setting and its members. This is like the dancer who reflects the dance and the *yogini*, or female yoga practitioner, who reflects inner growth and outer physical strength and endurance. Likewise, qualitative research methods are related to dance in another way, in that the body is the instrument of dance and the researcher is the research instrument in qualitative work. Furthermore, in yoga, the body and mind are integrated in all movement, work, and meditation in order to walk in balance of all phases of living.

■ Questions Suited to Qualitative Research Methods

1. Questions of the quality of a given innovation, program, or curriculum

2. Questions regarding meaning or interpretation

3. Questions related to sociolinguistic aspects of a setting

4. Questions related to the whole system, as in a classroom, school, school district, and so on

5. Questions regarding the political, economic, and social aspects of schooling or society

6. Questions regarding the hidden curriculum

7. Questions pertaining to the social context of schooling, such as race, class, and gender issues

8. Questions pertaining to implicit theories about how the social world works

9. Questions about a person's views on his or her life and work

10. Questions that are viewed as controversial

11. Questions related to public policy

■ Using Theory in Qualitative Research

As the reader is most likely aware, in qualitative work, theory is used at every step of the research process. Theoretical frames influence the questions we ask, the design of the study, the implementation of the study, and the way we interpret data. In addition, qualitative researchers develop theoretical models of what occurred in a study in order to explain their findings, warts and all. Qualitative researchers have an obligation to fully describe their theoretical postures at all stages of the research process, just as the choreographer fully describes and explains each component of a dance plan. As a choreographer, I was always looking for the asymmetrical movement in order to tell the story in some kind of symmetry. This is what I hope these experiences will help the reader to accomplish. Like the student of yoga, who works for unity in mind, breath, and body, the qualitative researcher ultimately is looking for this unity in the end in the final written report.

Speaking for myself, I have had many influences throughout my career, but the most notable influences from my own experience as a teacher include the work of John Dewey on education and art as experience, aesthetics, and habits of mind. In addition, I've looked to the work of Elliot Eisner on arts-based approaches to educational research and connoisseurship; critical pedagogy, cultural studies, and the work of Henry A. Giroux, Joe Kincheloe, and Shirley Steinberg; feminist theory and the work of Jane Flax and bell hooks; and postmodern sociology and the work of Norman Denzin. My former dance instructors and the texts written by my past teachers—Erick Hawkins and Merce Cunningham, Martha Graham and various teachers from her school, Margit Heskett and teachers from the Twyla Tharp school—have made a deep impression on my work. I call myself a critical, postmodern, interpretive interactionist with a feminist artistry. In addition, the works of Paolo Freire and Myles Horton have influenced my thinking. Horton's idea that we have our own solutions within us fits perfectly with interpretive work. Likewise, Freire's education for freedom is identical to Merce Cunningham's philosophy of dance. My own qualitative research projects have been guided by the theoretical frames of interpretive interactionism and critical pedagogy. This does not mean that other frameworks are incompatible or not useful. I have great affinity for phenomenology as described by Max Van Manen, the work of Valerie Yow on oral history, and the work being done in medicine by Atul Gawande in terms of narrative case studies. In qualitative research methods, of course Harry Wolcott, Norman Denzin, Yvonna Lincoln, Egon Guba, Judith Pressle, Irene and Herbert Rubin, and Valerie Polakow have influenced me in many ways.

A great deal of the debate over qualitative methods has to do with the issue of theory and its place in the research project. I would characterize this as a struggle of values. Many feel there is definitive knowledge about how to proceed in research substantively, theoretically, and procedurally. Others see research as a way to pursue moral, ethical, and political questions. I ask learners to think about and read many texts to get a feel for what resonates with them, what makes sense given their points of situation in doctoral studies. I often say that, in order to do qualitative research, you must accept that there is no universal truth, that all findings are tentative and context based, and that we live in an irrational and chaotic world. For those who want a neat and tidy social world, this is definitely not for you. We begin with a serious and prevailing curiosity about the social world. The focus is on formulating good questions. Many scientists, artists, and writers agree that formulating a question and problem is the essence of creative work. As Einstein and Infeld (1938) said:

> The formulation of a problem is often more essential than its solution, which may be merely a matter of . . . skill. To raise new questions, new possibilities, to regard old questions from a new angle, requires creative imagination and marks real advances in science. (p. 35)

What this means for the purposes of this book and the study of qualitative research methods is that we transform ourselves by looking and seeing what is before us in our observations. In addition, we hear the data as it is spoken to us in interviews. We refine our narrative writing skills to be able to put forth a trustworthy, credible, and authentic story.

The world for the qualitative researcher is tentative, problematic, ever changing, irrational, and yes, even chaotic. Many qualitative researchers see research as participatory, dialogic, transformative, and educative. It may be a constructivist, critical, and transformative approach to research. In this text, I am using the metaphor of stretching in dance and yoga as an art form to illuminate some of the many components of qualitative work—observation skills, interview skills, and the role of the researcher skills—in order to arrive at that level of participation, transformation, and education.

■ Artistic Approaches to Qualitative Research

There is a long and embedded theoretical history for me starting with John Dewey's (1958) *Art as Experience* (*AAE*). In fact, for Dewey, art was not about daydreaming but about providing a sense of the whole of something, much the same way qualitative researchers see the whole picture in their slice of the case under study. In *The Early Works* (*TEW*), Dewey (1967) states:

> The poet not only detects subtler analogies than other (men), and provides the subtler link of identity where others see confusion and

difference, but the form of his expression, his language; images, etc. are controlled by deeper unities . . . of feeling. The objects, ideas, connected are perhaps remote from each other to the intellect, but feeling fuses them. Unity of feeling gives artistic unity, wholeness of effect, to the composition. (p. 96)

So for Dewey, especially in *TEW,* imagination was highly valued and was explained in terms of feeling. In 1931, when Dewey delivered the first of the William James lectures at Harvard, the subject was that which became *AAE.* He was roundly criticized at the time. Most problematic was Dewey's suggestion that art is about communication and experience. How history changes us! It would be very difficult today to find someone to disagree with the wisdom of Dewey on this point. Dewey refused to separate art from ordinary experience. He said that the artist should "restore continuity between the refined and intensified forms of experience that are works of art and the everyday events, doings, and sufferings that are universally recognized to constitute experience" (Dewey, 1958, p. 97). The qualitative researcher is involved in this artistic activity because he or she must describe and explain the lived experience of participants in his or her study. I have written (2008) on the clarity of art as experience and that we are always looking for that which is beyond the obvious. That is the artist's way, the dancer and choreographer's way, and the way of the qualitative researcher. Furthermore, the notion of developing habits of mind that Dewey spoke of and wrote of in 1938 still resonates for me today. Although he was writing about logic, the main idea of making your mind work is important. It pertains to the dancer and the yogi for mind and body union. In fact that word, *yoga,* means *union* and is interpreted as union of mind and body.

I ask the reader's indulgence here as I try to make the ordinary activities described in this text evocative of Dewey's notions in order for the prospective qualitative researcher to eventually become aware of a critical approach to art and research as experience. This is the only way that makes sense for me. But even those who may not wish to revisit Dewey's ideas on art as experience or on habits of mind can certainly find points of integration and connection in whatever other theoretical framework is used to guide their research.

For the purposes of this text, the description and observation exercises relate to the field-focused nature of the work and to developing the habit of observing, seeing, and writing about that experience. The role of the researcher exercises relate to the self as research instrument and to developing the creative habit. The interview exercises relate to the interpretive, expressive nature of this work, the presence of voice in the text, and developing this habit and writing about it. The analysis of data exercises and Internet inquiry exercises relate to the points of coherence, insight, and utility. As an umbrella for all this, narrative writing is the key to communicating purposes, methods, findings, and interpretation of the study and developing the habit of using the brain for analysis and interpretation. In addition, any use of poetry found in the interview transcripts or documents also fall in this frame. In fact, photography, video, or any web-based inquiry can also fall into the creative habit and analytical habit of mind. I ask all my students to be above average in writing ability, among other things, in order to complete a qualitative dissertation and subsequent qualitative work. The arduous habit of writing of course demands critical thinking. Here again, the researcher as research instrument must be actively involved in developing insight, creativity, acute observation, sensitive interviewing, and all digital techniques to tell the story of what was found in the research project. This can most often be achieved through diligent practice.

■ Why Try These Exercises?

Although these exercises are for those who are interested, they are not for everyone. Not everyone has tapped into his or her artistic intelligence. I am in agreement with those writers who have found that we all have an artistic side as well as multiple intelligences (see the work of Howard Gardner), but I have also found that the learner must take an *active* role in discovering his or her artistic intelligence. I am constantly amazed at learners who tell me that they cannot draw. When we do the drawing exercise in class, they are doubly amazed that, to some extent, they can draw, and when they revisit that exercise later, they have to admit that they can indeed draw. More on that topic will follow when we arrive at that exercise. In addition to this active stance, the artistic theoretical frame that drives these exercises is critical, transformative, educative, and ethical. These exercises were created from a lifetime and career of reading and action. More or less, these writers were part of my dialogue with myself: John Dewey, Elliot Eisner, Myles Horton, Henry Giroux, Maxine Greene, Joe Kincheloe, and Paolo Freire in education. Twyla Tharp, Erick Hawkins, Merce Cunningham, and Martha Graham have influenced my thinking about dance and choreography. Yoga writers who influenced me are both contemporary, like Rodney Yee, and historical, like Yogi Sachitananda. Just as the dancer might look before leaping, so the reader of this text might look with a critical and enlightened eye.

I must ask the reader to do something that is very difficult: Give up one view of the world and imagine another.

In other words, just because there are exercises for all levels, this does not mean that in the final product, that is, the creation of a work of art and science, it will be easy. It takes persistence, determination, preparation, passion, diligence, and above-average writing skills to do all this. Nevertheless, I purposely chose to write in a clear, descriptive, narrative style. I write in ordinary language and in my voice for the following reasons:

1. To *disrupt* what some have called academic writing, which distances the reader from what is written and denigrates the reader's experience. Academic writing often excludes many people who want to be part of the conversation. By eliminating thick jargon and tired phrases, we open the space for a creative use of ordinary language. Thus, more people may read our stories.

2. To *educate and engage* the reader, who may not, up to now, have had an interest in qualitative research. I am writing both for people who love qualitative research and those who think they hate qualitative research. In addition to this, I like to educate my coworkers. I often tell my students that they must show all the texts they have read to the professors who sit on their committees if for no other reason than to let those members know of the vast array of written texts about qualitative research. Another point of educating is to make the learner aware of the amount of time, money, and energy involved in the undertaking of a qualitative study. (See Appendix J for the recent documentation of the cost of a recently completed qualitative dissertation for example.)

3. To *inspire* the reader to go further and read the writings on theory and practice in qualitative research. Students who are turned on to qualitative research purchase all the texts referred to in whatever they are reading on the topic at the moment. They also join discussion groups and LISTSERVs, and they scan dissertation abstracts regularly for the latest qualitative dissertations. Later in this text, you will find numerous examples of web-based resources to inspire further study. In fact, since the second edition of this book, Internet inquiry has evolved into a mainstream research genre in many fields, and that will be discussed later in this third edition.

4. To *demystify* the research process by the use of ordinary language, thereby opening up the pool of researchers in our field. *In a very real sense, qualitative research is contributing to a more democratic space for doing research.*

As in the past, currently, the language of quantitative research is off-putting to many students, and as a result, educational research at least has been disavowed or overlooked. For too long, research has been cloaked in secrecy and jargon, and behavior to keep things hidden has been the rule. We call it the curse of the ivory tower. In addition, when research reports are published with all the jargon and formulas, there is a distancing of the reader from the report. Here, in this paradigm, the reader is part of the report because it is made understandable. With the questions of the post-modern era forever before us, we have no choice but to deconstruct and demystify the research process. One way that qualitative researchers are ahead here is that we already value and use ordinary language. We already value ordinary people in ordinary life. We want to describe and explain the social world. Some researchers even go so far as to hope for a better world through application of research findings. We are researchers of subjectivity and proud of it.

5. To *democratize* the research process. Qualitative research techniques open up the process of research to many more researchers who take responsibility for the rigor and high standards of this work. Consequently, there is less emphasis on only a few elite individuals taking ownership of these approaches. In this postmodern era of constructivist models of learning and teaching, which open up knowledge acquisition, qualitative research is a part of opening up all research processes to those previously excluded from the conversation. The Internet and all its strengths and weaknesses have given us a new and bold democratic path. The wiki world, for example, invites the viewer to add content with verification to a knowledge base, such as Wikipedia. Our job is to be critical agents who tune our bodies and minds to adhere to the rigor of qualitative research methods and to disseminate the research findings with care, with precision, with ethical awareness, and with authenticity. This is in short taking part in the democracy of the research process.

■ How to Use This Book

The whole idea of this text is to get the reader to stretch. These exercises should get you started in the actual experience of doing observations and conducting interviews. After all, fieldwork is mostly work, so you must strengthen your body and spirit to do it well. You may wish to scan the book before beginning the exercises. I begin with observation exercises in order to force you into another way of thinking about and seeing the world. Think of these exercises as making you a stronger, more flexible, and more fluid researcher, just as the dancer becomes stronger, more flexible, and more fluid after stretching. Likewise, the practice of yoga, if done correctly, can strengthen body, mind, and spirit. There is a regularity and discipline to fieldwork, much like that of dance and yoga, and these exercises are progressive in difficulty in each of the major sections.

The exercises provide a process for developing skills in the main techniques of qualitative research methods, that is, observation and interview. The shape of these exercises developed over time and will continue to develop. In dance, there are no static points. Likewise, in qualitative research, there are no static points, only reshaping the approach and continually questioning and analyzing. The reader may also notice that many of these exercises refer to the arts in order to broaden the conversation and thinking about the research process. I have always found that my own background in drawing, photography, drama, and dance has provided the foundation for activities that eventually provide access for many students to improve as interviewers and observers of the world. For me, research is alive and active. It is the most exciting use of many ways of looking at and interpreting the world. I hope these exercises will convey a portion of that enthusiasm for knowing. By the way, enthusiasm and passion for your research project is essential to sustain you in the many hours of labor in the field. Fieldwork is indeed hard and demanding work after all is said and done. This work needs to be documented, and a favorite and emancipating way to do this is through writing a reflective journal while reading this book and doing the exercises. Starting in the next chapter, and at the end of every chapter, you will be asked to compile entries through various exercises in journal writing. Thus, after reading and working through the exercises you select, you will have your research reflective journal started as a test drive, so to speak, for your actual dissertation or other research project.

■ The Audience for This Book

This text is for anyone who wants to practice the two most prominent techniques used in qualitative research projects: interviewing and observation. At the same time, writing skills and digital inquiry skills go hand in hand with these two cornerstones of our work. In addition, some exercises included in the text are for the purpose of developing a stronger awareness of the role of the researcher through writing, reflecting upon ethical issues in qualitative work, and developing the creative habit. I have become aware, since the first edition of this text, that it is also for teacher-educators and teachers in training. Because the field of teacher education relies on action research, critical case studies, and teacher research, this book may assist those members of the educational community. As an educator, I would include the group of students of research as a major portion of the audience. Colleagues who wish to practice qualitative research methods are certainly included as well. May the reader of this text have a passion for disciplined inquiry, a high tolerance for ambiguity, a rich imagination, an open mind toward ordinary language usage, and a very good sense of humor.

■ Bibliography as Resource List

The following references are resources to help you navigate this text and learn more about qualitative research. They have been remarkably helpful to students.

Berg, B. (2007). *Qualitative research for the social sciences* (6th ed.). Boston: Allyn & Bacon.

Best, J. (2004). *More damned lies and statistics: How numbers confuse public issues.* Berkeley: University of California Press.

Cole, A., & Knowles, G. (2001). *Lives in context: The art of life history research.* Sherman Oaks, CA: AltaMira.

Charmaz, K. (2000). Grounded theory: Objectivist and constructivist methods. In N. K. Denzin & Y. S. Lincoln (Eds.), *Handbook of qualitative research* (pp. 509–535). Thousand Oaks, CA: Sage.

Creswell, J. W. (2007). *Qualitative inquiry and research design: Choosing among five approaches.* Thousand Oaks, CA: Sage.

Denzin, N. K., & Lincoln, Y. S. (2002a). *Handbook of qualitative research* (2nd ed.). Thousand Oaks, CA: Sage.

Denzin, N. K., & Lincoln, Y. S. (2002b). *The qualitative inquiry reader.* Thousand Oaks, CA: Sage.

Denzin, N. K., & Lincoln, Y. S. (2003a). *The landscape of qualitative research: Theories and issues.* Thousand Oaks, CA: Sage.

Denzin, N. K., & Lincoln, Y. S. (2003b). *The strategies of qualitative inquiry* (2nd ed.). Thousand Oaks, CA: Sage.

Denzin N. K., & Lincoln, Y. S. (2006). *Handbook of qualitative research* (3rd ed.). Thousand Oaks, CA: Sage.

Dewey, J. (1938). *Experience and education.* New York: Collier.

Dewey, J. (1958). *Art as experience.* New York: Capricorn.

Dewey, J. (1967). *The early works.* Carbondale: Southern Illinois University Press.

Dochartaigh, N. O. (2002). *Doing research on the Internet: A practical guide for students and researchers in the social sciences.* Thousand Oaks, CA: Sage.

Eisner, E. W. (1991). *The enlightened eye.* New York: Macmillan.

Estes, C. P. (1992/1996). *Women who run with the wolves: Myths and stories of the wild woman archetype.* New York: Ballantine.

Flick, U. (2002). *An introduction to qualitative research* (2nd ed.). Thousand Oaks, CA: Sage.

Gawande, A. (2007). *Better: A surgeon's notes on performance.* New York: Metropolitan Books.

Freire, P. (2007). *Pedagogy of the oppressed, 30th anniversary ed.* New York: Continuum.

Janesick, V. J. (1994). The dance of qualitative research design: Metaphor, methodolatry, and meaning. In N. K. Denzin & Y. S. Lincoln (Eds.), *Handbook of qualitative research* (pp. 209–219). Thousand Oaks, CA: Sage.

Janesick, V. J. (1998). *"Stretching" exercises for qualitative researchers.* Thousand Oaks, CA: Sage.

Janesick, V. J. (1999). A journal about journal writing as a qualitative research technique: History, issues, and reflections. *Qualitative Inquiry, 5*(4), 505–523.

Janesick, V. J. (2000). The choreography of qualitative research design: Minuets, improvisations, and crystallization. In N. K. Denzin & Y. S. Lincoln (Eds.), *Handbook of qualitative research* (2nd ed., pp. 370–399). Thousand Oaks, CA: Sage.

Janesick, V. J. (2007). Oral history as a social justice project: Issues for the qualitative researcher. *The Qualitative Report, 12*(1), 111–121. Retrieved December 1, 2009, from http://www.nova.edu/ssss/QR/QR12-1/janesick.pdf

Janesick, V. J. (2008). Art and experience: Lessons learned from Dewey and Hawkins. In J. G. Knowles & A. L. Cole (Eds.). *Handbook of arts in qualitative inquiry: Perspectives, methodologies, examples and issues* (pp. 477–483). Thousand Oaks, CA: Sage Publications.

Knowles, J. G., & Cole, A. L. (Eds.). (2008). *Handbook of the arts in qualitative research.* Thousand Oaks, CA: Sage.

Locke, L. F., Spirduso, W. W., & Silverman, S. J. (2007). *Proposals that work: A guide for dissertations and grant proposals* (5th ed.). Thousand Oaks, CA: Sage.

Markham, A. N., & Baym, N. K. (2009). *Internet inquiry: Conversations about method.* Thousand Oaks, CA: Sage.

Merriam, S. B. (2009). *Qualitative research: A guide to design and implementation* (Rev. ed.) San Francisco: Jossey-Bass.

Piantanida, M., & Garman, N. B. (2009). *The qualitative dissertation: A guide for students and faculty* (2nd ed.). Thousand Oaks, CA: Corwin.

Rossman, G. B., & Rallis, S. F. (2003). *Learning in the field: An introduction to qualitative research* (2nd ed.). Thousand Oaks, CA: Sage.

Rubin, H. J., & Rubin, I. S. (2005). *Qualitative interviewing: The art of hearing data* (2nd ed.). Thousand Oaks, CA: Sage.

Schwandt, T. A. (2007). *The SAGE dictionary of qualitative inquiry.* Thousand Oaks, CA: Sage.

Smulyan, L. (2000). *Balancing acts: Women principals at work.* Albany, NY: SUNY Press.

Wolcott, H. (1994). *Transforming qualitative data.* Thousand Oaks, CA: Sage.

Wolcott, H. (1995). *The art of fieldwork.* Walnut Creek, CA: AltaMira.

Wolcott, H. F. (2002). *Writing up qualitative research* (2nd ed.). Thousand Oaks, CA: Sage.

CHAPTER 2

The Observation and Writing Habit ■

We cannot create observers by saying "observe" but by giving them the power and the means for this observation and these means are procured through education of the senses.

—Maria Montessori, Italian physician and educator, 1870–1952
(Early Childhood Education Today, *Morrison, 2008*)

When Maria Montessori wrote these words, they caused quite a commotion in her field. It reminds me of one of my favorite dance teachers in New York City, from the Merce Cunningham School at Westbeth, who once asked all of us in class to observe her movement closely. The reason to observe so carefully, she said, "was to become more aware of your own body and mind" and to "internalize" the movement. She emphasized that until we could observe ourselves and each other, we would not be able to dance with freedom. From this, I began to learn that observing carefully was so focused an activity that, in order to teach others to observe, I needed a way to introduce observation practices that allow the learner to develop this habit of mind and develop narrative writing skills.

Similarly, in a recent yoga class, my teacher demonstrated a variation on a headstand that prepares one for a headstand. First, the teacher demonstrated it to give an idea of what it should look like. Then, he asked us to write down in our yoga practice journals all that we had observed. Next, we read them to each other and the class before we started practicing the pose. We started out with one step, and moved to the next, and moved to the next. The exercises described in this section start out simply and grow in complexity as the learner becomes more practiced in each activity. Again, this is not meant to be a slavish following of a recipe. Each person may improvise at any given point so that he or she continues to claim an active part in the activity. One size does not fit all.

I begin with an exercise taught to me by my favorite high school art teacher, which is simply to observe a group of objects. Although this exercise was meant originally for drawing, I find it helpful when introducing prospective researchers to the activity of description. Also, this exercise may help to place observation in a historical context as a step in an ancient and continuing journey. One aim I have for all of my work is to understand qualitative inquiry in a historical context that goes back ages in order to recognize a common history. I like to frame the history of observation of one's environment from an art history perspective that begins in 3000 BCE, when the Chinese master painters began recording their observations of everyday moments in their environment with descriptions of trees, orchids, rocks, plants, water, and so on. I start arbitrarily with the Chinese because of their long history of appreciation for nature, observation, and aesthetics and because of their systematic and methodical approach to documentation. Likewise, I stress the meaning of the term *empirical* at this point to reflect its meaning, "relying on direct experience and observation," as the cornerstone of qualitative work. It is confounding to me that many students think the word empirical is a synonym for statistics or numbers. It takes a while for them to realize that numbers are one, two, or three times removed from the actual experience being described! In addition, I ask students to find the definition for the word empirical in three different dictionaries. This gets them thinking in a new way. In this first exercise, we start with a still life scene because it grounds learners in observation, and it is less demanding than observing and describing people. While every single detail seems impossible to document, learners are advised to get as many details into their descriptions as possible.

Exercise 2.1   ■

Observing a Still Life Scene   ■

Purpose: To observe and describe an assortment of objects on a table in 5-minute increments twice.

Problem: To see these objects from your position in the room.

Time: Take 10 to 20 minutes of observation time and another 10 to 20 minutes of reading to the group or to someone what you wrote to get feedback and try again.

Activity: Set up a table in the center of the room so that viewers will have multiple positions and vantage points as they observe at least five objects, each of a different shape, texture, size, and color. Select any objects you wish. For example, I recently arranged the following:

1. A coffee mug with various photos on it encased in a plastic outer shell with a circular, fingernail-size Starbucks logo
2. A 5-inch by 7-inch, framed, empty, stand-up photo holder
3. A deep purple steel water bottle
4. A pile of newspapers approximately 4 inches high
5. A *Webster's Dictionary* in a red, textured cover

I placed the items listed above on a portable table for all to view from their seats. I envision a room of worktables arranged in a rectangle so that each person has a direct view of some portion of the scene. If the reader is working on these exercises alone, the reader selects a space for this activity. Space and how it is used are as critical for the qualitative researcher as they are for the dancer. The second time around with the description, try moving yourself to

(Continued)

(Continued)

another vantage point to get a sense of the importance of the placement of the observer.

Aim: In the time allotted, describe what you see on the table. Use descriptive terms and field note format. Although there are countless ways to take field notes, for the purposes of the group, we all agree to use one format that includes space for the researchers to write notes to themselves on the left third of the page and to write the actual descriptive field notes on the right two thirds of the page. In the upper right-hand corner, there is space for the following information: date, time, place, and participants. Learners are urged to develop a system of pagination and classification of notes. For example, for descriptions of settings, one may use a color code at the top of the page or select a different color of paper. For descriptions of interactions in a social setting, another color code at the top of the paper may be used and so on. In any event, the object is to encourage learners to create their own systems of coding that work for them. Figure 2.1 shows a sample of a field note page in the format and style we have adopted because it is effective for our purposes. My students and I find that this format is helpful because half of it is for the field notes and half of it is for notes to ourselves.

Figure 2.1   A Sample Field Note Format

Field Notes

By Ruth Slotnick (2010), University of South Florida (USF)

Date: January 27, 2010

Time: 10:17–11:17 a.m.

Location: University Medical Clinic

Type: Nonparticipant observation

USF Health Dermatology and Cutaneous Surgery

Notes	Observations
	10:17 a.m.
	Getting To:
I pretend to look at my watch because I feel an older man observing me. To help blend in, I glance at the time to appear to be less of a stranger.

Need to look up glass barriers on the web. | I am sitting in cubical-like fishbowl located between sections A and C of the university medical health clinic. From the main entrance of the clinic, one can get to this space by walking straight ahead through the automatic doubled-glass doors and proceeding along the industrial-carpeted, fluorescent-lit corridor passing the Quest Diagnostics waiting room on the right. Also on right and left sides of the wallpapered hallway, there are a series of 45-inch-by-39-inch reproduced watercolor paintings depicting Florida sea and landscapes. When the hallway comes to a dead end, make a left and continue down corridor until intersecting an adjoining hallway. Make a hard right and walk past patient check-in desks A and B on the right and left, respectively. |
| | **Just Outside the Clinic:** |
| How many patients frequent the medical clinic each year? | To get to the clinic, immediately upon passing patient check-in station A, make a hard left into the waiting room outlined by two 15-foot-by-20-foot, separately placed and freely standing, floor-to-ceiling, C-shaped transparent glass walls with frosted, horizontal bands ranging from 2 to 3 inches running through the middle. The glass structure is framed by a dark brown plastic molding that outlines the tops and bottoms. The vertical edges are lined with a 1-inch stainless steel, silver-matt border. |

(Continued)

(Continued)

Notes	Observations
	Entering
	10:33 a.m.
	Advancing through the opening of the two C-shaped glass dividers, one passes a large 16-inch diameter, oxblood-red, fishbowl-shaped, mass-manufactured, low-fired planter with a wide-mouth flange. The ceramic pot sits on top of a plastic, faux-Japanese, footed plant stand. The plant is not real. It is an artificial, leafy palm tree plant with three perfectly placed stalks, one high and to the right, one low to the left, and one in
Need to look up the type of footer for ceramic form.	the middle. To the right and to the left of each glass wall are two sets of high-density, foam-filled, heavy-duty triplet plastic seats (each with a separate cushion and backrest) mounted on top of a tubular steel frame that is a matted blush color. The metal framing is a glossy rose. The sets of metal chairs are joined together by a half-moon-shaped table that stands about 2 feet from the floor. This surface is made of a wood composite laminate, which is also the color of dusty rose. A clear plastic pamphlet holder is stacked four deep
Note to self: Older patient signals to me that he cannot hear the nurse when she calls out his name.	with tri-fold pamphlets. One pamphlet advertises the "Cosmetic and Laser Center" on 60-lb., glossy white paper. A second pamphlet is labeled "Your Dermatologist." The color for this brochure is benthic blue with white Times font letting. In the center of the half-moon-shaped coffee table is another 8.5-inch-by-11-inch, tri-stacked, clear plastic brochure holder. The bottommost row holds a multicolor, oversized bookmark with a five-step guide on how to examine your body for melanomas. Each step is allotted its own space with a gesture drawing of how to examine your body. Two other pamphlets

Notes	Observations

remain in the stack: a placemat-like, double-sided sheet describing how to examine your skin and also how to improve tone and treat sun-damaged skin. Another one-page, high-gloss, professionally printed, full-sized brochure shows how skin cancer is "treatable and beatable with early detection." There are eight steps given on how to protect and prevent skin cancer.

Two spiral-bound half notebooks adorn the middle of the table. A laser copy color print is placed inside the front sleeve and displays a photo of a male doctor. He appears to be in his mid-50s. Gray hair and mustache. Friendly smile. He is wearing his doctor's overcoat with a sky-blue, button-collared shirt and colorful, silk tie. A pair of reading glasses hangs around his neck. His name is Dr. Sam Fiske, and he is both the professor and chairman of the Dermatology and Cutaneous Surgery unit as well as the local newspaper correspondent. The caption written over Dr. Fiske's image reads "Ask the Expert." The university logo is prominently visible inside each notebook as are various newspaper and magazine articles on skin cancer, acne, how to remove moles, rejuvenate skin, remedy facial veins, smooth flaky feet, and combat fungal infections. Other articles include such topics as Vitamin D, eczema, and premature skin aging, useful information on how too much sunscreen can be harmful. All these articles have been written by Dr. Fiske. A final brochure, also a tri-fold, is advertising "Humira," which is used to treat psoriasis. This brochure includes a hologram that shows a before and after effect of using the drug.

(Continued)

(Continued)

Notes	Observations
	Left Side of Glass Cubicle
	11:05 a.m.
	Two sets of seats are in the middle, placed back-to-back in a set of three and a set of two. Inside this cluster of seats is a white cubicle containing a yellow-pages phone book and a black, touch-tone office phone. A placard above reads "Please Limit Telephone Calls to 5 Minutes."
I wonder if elementary school children are covering the topic of the sun in their science classes.	A gray-colored fabric cubicle about 8 feet high, spanning three panels long for a grand total of 14 feet, has a 10-foot-by-4-foot white poster created by a local elementary school. The word *shade* is spelled out in big, black, block lowercase letters. About 500 names surround the word.

Date: January 28, 2010

Time: 8:47–9:47 a.m.

Location: USF Medical Clinic

Type: Nonparticipant observation

USF Health Dermatology and Cutaneous Surgery

Notes	Observations
	8:47 a.m.
How do people know what clinic they are supposed to go to or sit in? The clinic waiting rooms are not clearly labeled beyond a letter A, B, or C, etc.	I am sitting in the waiting room of clinic A. Not a single patient is waiting. Clinic D check-in desk is busier. At D, there appear to be a few individuals in line. A heavyset Black female in her 40s wearing jet-black sweatpants, a charcoal-gray T-shirt, a hooded, zip-up, black jersey sweatshirt, and a cobalt-blue, loosely woven knitted cap converses about a missed appointment with the receptionist.

Notes	Observations
Appears to be quite a few Black women working in the receptionist area. I notice people wearing denim jeans and sweatshirts. Does this mean that people are either too sick or not interested in getting dressed up for the doctor?	A White man in his 30s wearing a pair of faded denim jeans, puma sneakers, and a chocolate brown-colored, hooded sweatshirt drinks a can of ginger ale or Sprite, sniffling and sneezing while waiting his turn in line. He rocks from side to side. A receptionist from section C (across from D) escorts "Ms. Miller," a Black woman in her 40s, into the neighboring clinic waiting room. Still, no patients are sitting in the Derm Clinic area of A.
	8:56 a.m.
	Today, I notice the end tables have an assortment of popular magazines including *People, Newsweek*, and *Reader's Digest* placed on them. Another pamphlet has been added to the pile. It is pink brochure titled "Focus on Females."
	9:00 a.m.
Who is the patient?	First couple enters the waiting room. After checking in, they place a medical file in P. Weinberg's mail slot, which is affixed to the brick wall next to the clinic door. A nurse quickly opens the door and retrieves the file. The couple is in their late 70s. He is dressed in a pair of dark blue chinos, is 5 feet 9 inches, and is wearing brown Dexter shoes, white nylon socks, and a blue denim, button-down shirt. He has a heavy gold watch on his left wrist. He is holding his head in his left hand, which is slightly cocked to the left side, with his elbow resting on the armrest. He wears a pair of wire-rimmed prescription glasses. His hair is salt and pepper. The woman sitting to his right is wearing a winter, reversible, black and

(Continued)

(Continued)

Notes	Observations
	red down coat. The nurse pops her head out of the clinic door and calls, "James." The couple stands up and goes into the clinic with James leading.
This couple seems to be low income.	Moments later, the next couple sits down exactly where the last couple was seated. The woman speaks to the man, "You got all my papers?" They place their paperwork in Dr. Weinberg's mail slot. The woman is rotund. She is wearing a stonewashed denim jacket with a small pocket over the elbow. The woman appears to be in her mid-50s. She also leans, but to the right. She appears to be resting or catching her breath. Maybe she is listening to a message on her cell phone? Her back is to me. Can't tell. Suddenly, she turns her head left toward the clinic door. She is not using her cell phone! Her partner, who has gone to the men's rest room, returns, scuffing his black leather sneakers against the carpet as he walks. He is wearing a red and black-checked flannel shirt. His dirty blond, unwashed hair, which is shoulder length, is tucked under a gray-blue baseball cap. He is also wearing prescription glasses and a very worn pair of denim jeans. He has a mustache. I cannot tell if he is shaven. She is mumbling something like, "We can't go downtown . . ." when her name "Leslie" is called. She continues this thought in breathy tones, "Try to take that lady those papers?" She is extremely bottom heavy and is wearing a pair of dark brown sweatpants with two lime green vertical stripes running down the sides. As she enters the clinic, she slowly shuffles and waddles painfully, teeter-tottering down the clinic hallway behind her nurse and her partner.
Something smells? These people perhaps? Not bathed? Clothes unwashed?	

Notes	Observations
	Another man arrives; he is in his late 40s. Dark skinned. Maybe Hispanic? Wearing a pair of dungarees and a light gray-colored sweatshirt with an untucked T-shirt hanging out the back of his pants.
At first I thought this was a lesbian couple. The son was so obese that he had breasts. His hair was cropped very short. I only realized that he was male when he disappeared to the men's room after his visit. Earlier, a man had been directed to the men's room so I knew where it was.

The man seems to be high-functioning autistic. He is clean but overweight. Slurred speech. Some kind of speech impediment. Has a library book tucked under his arm so is therefore literate. Mother also comes with a novel. | A female arrives. Also for Dr. Weinberg. She smells like a cigarette smoker. This woman is in her 70s. Directly behind her, another female in her 40s—very large and out of breath with short, buzzed-cropped brown hair—sits down next to her. She is having trouble sitting down due to her large body size. Loudly, she whispers in slurred speech to the older woman, "No wonder why! Pants are twisted." She has on a pair of slip-on, black leather shoes with Velcro straps, and cotton, white socks are on her feet. Her pelvis area is very full and extended. Suddenly, the clinic door opens, and her name is called. In her surprise of being called so soon, she bolts toward the clinic nurse, running—all 200 pounds of her. The older woman (maybe her mother?) is following quickly behind. They exit at 9:45 a.m. The older woman, wearing a pair of dungarees and a light blue cotton sweater, goes to check out. The larger woman goes to bathroom around the corner. Wait, that's the men's rest room. Maybe she is not a she? Maybe he? Yes, male! |
| So far all the patients are coming for Dr. Weinberg. The mother and daughter couple is clearly dressed up for the doctor. The mother has an Eastern European accent. | Next patient: A well-dressed older woman in her 80s enters the clinic with a woman in her 50s or 60s escorting her. The impeccably dressed elderly woman is about 5 feet with cropped-short silver hair, wearing a matching pants and top outfit. Her osteoporosis makes her stoop, causing her pants to rise slightly above her ankles when she walks. |

(Continued)

(Continued)

Notes	Observations
	The blond woman escorting her is also well dressed. She holds onto the older woman's arm as they exit the clinic. They had come in earlier in the same fashion but did not appear to wait. The nurse opened the door to the clinic, and they were simply ushered in with the nurse commenting, "We were just looking for you." In response, the older woman stated, "It's like you were waiting for us! Such service."
Based on the headscarf and seeing no hair, I wonder if this woman is undergoing chemo.	An older female in her 70s enters the waiting room, wearing a sandy-colored headscarf and a suede jacket of the same color and black slacks with a row of buttons tapering down to her ankle. She has on a pair of slip-on black flats. Just when she sits down to fill out paperwork on a clipboard, she is called in. Her glasses are black, square-framed bifocals. "Petal," the nurse beckons.
For some reason, I found myself bored by this description.	An army vet in his 70s arrives in the space. He is wearing faded fatigues, an ochre-colored rain jacket, and a sandy-colored, button-down shirt with his glasses case (brown) placed in a pocket over his heart. He is wearing a pair of slip-on leather Bass shoes. The tops appear worn, but the soles are in good shape. He is clean shaven. Tufts of white hair spill out from under his marine army cap with a marine seal embroidered on the brim. The word *marine* is sewn on with golden thread. His socks are cottony white. He steps up to leave and greets the Hispanic man who entered earlier in a businesslike manner. The very first couple, 9:37 a.m., exits too.
Lots of women using the women's bathroom. No public-use bathrooms in the clinics?	The woman with the headscarf exits the clinic to use the women's bathroom, walking briskly for her slight frame. Her large, pouch-like, soft, matt-black leather bag is slung over her left shoulder.

Notes	Observations
	It has golden buckles at the corners. Her bag is now slung over the right shoulder on her way back.
Exit space, 9:47 a.m.	9:43 a.m.
	The waiting room is empty. Only one older White female with black hair signing in at station A.

In this exercise to observe objects, the learners take notes as the observation takes place. After 10 minutes, they change seats and observe and describe the objects for another period of 10 minutes. The purpose for doing this is to experience the change of seating and viewpoint for description. Later, we discuss the reactions of the students. An interesting result of this, for example, is that one person may simply write something like, "there is a vase of flowers on the table," whereas another may describe that same vase of flowers with the precise Latin names for each flower; the type, shape, and size of the glass; and the tints and hues seen in the vase. Learners share their field notes with class members and get feedback from the group members as well as myself. We then repeat the entire process. When taking field notes for description, be sure to write notes to yourself as reminders and memos. See this fine example of field notes by Ruth Slotnick (2010).

■ Description of a Still Life Scene: 10 Minutes' Worth

I begin with the coffee mug. The mug appears to be approximately 8 inches tall, made of plastic that covers various travel photos. It is shades of beige, brown, blue, and red. It stands with a black plastic lid with a flip-up flap that allows for drinking or stays closed, most likely to keep the coffee hot. The photos, which I call travel photos, are scenes about the size of a postage stamp. One scene is at the beach, another is a sailboat alone in a cove near a rock formation. Next are three photos of children. One is a giggling child approximately 3 years of age, wearing wide-brimmed white sunglasses, near a sandy shore. The photo is too small to identify the lake or body of water. The next photo has four children, toddlers, and one adult sitting on what appears to be a picnic blanket. The next photo is a young boy and girl, heads together, smiling, approximately 7 or 8 years of age. About a half inch below the screw-top lid is a rectangular box with the words in script saying, "Create your own tumbler." I can only see from where I am sitting, just to the right of the mug approximately 12 inches away, the words, "Twist off the base and remove this template." I will return and review this later as my time is up.

Beginning the research reflective journal and discussion:

How did you approach this exercise in the first 5 minutes? The second 5 minutes? Did you do anything differently in the second observation after you changed seating? What did you learn from the feedback from the class? Following this discussion, write about this in your newly created researcher reflective journal.

What was most difficult for you in this exercise?

Find a good notebook or set of notebooks to accomplish this performance task. You may wish to have a separate notebook on observations, one on interviews, one for poetry, and one for photography and other categories. Or you may wish to keep your journal continuously, integrating all of the above in various eventual volumes. By the end of any set of these exercises, you will have a written product that traces your thoughts and feelings, memories and dreams, stories and partial stories. Continuing on, write at least 10 minutes on each of these questions. Each day, you will see yourself wanting to write more. Feel free to backtrack and edit, add to, or delete from your researcher reflective journal.

Rationale:

The questions are designed to discover how learners approach the task and how others approach the same task. Some begin by describing one object at a time from left to right, forgetting to situate themselves in the room. Others begin by describing the room and working their way to the table and then to the objects. By sharing their personal experiences, class members begin to see the wide variety of approaches to the task. They also see which members are skilled at description and which need to practice more of the art of description. There is no better teacher than example, so in class, when a skilled observer and describer reads a powerful description to the class, the lesson sinks in. After all, the researcher is the research instrument, and these practice exercises in observation and description are designed to bring home this very point.

Evaluation:

In addition to my evaluation of individual work, I ask the students to evaluate themselves. They are asked to fill out the following form, shown in Figure 2.2, or to write me a letter and return to discuss it at the beginning of the next class before starting the next exercise, the observation of the classroom setting. For those working alone, write a letter to someone you trust evaluating this process. I use this form to facilitate an active agency as evaluators of one's work. The questions are meant to be a heuristic tool and a starting point for the evaluative process. Most often, I ask learners to keep a journal of all their thoughts related to the class and these activities. Although many are intimidated by writing a journal, this can be a good evaluation tool and a valuable historical document for charting the role of the researcher in a given project. Journal writing is my first choice and suggestion for self-reflection and evaluation. Of course, the point is to allow the individual to become a stakeholder in self-evaluation.

Figure 2.2 A Sample Formative Evaluation

Evaluation of: (Name the exercise)

Date:

    a. Things I learned about myself as I described the objects under observation:

    b. Things I need to continue to work on for the next observation:

    c. How I describe my progress:

This baptism into self-evaluation continues with each practice exercise and is part of another overall system of self-evaluation, which looks like Figure 2.3. In addition, students begin to build their own portfolios and integrate these evaluations into researcher reflective journals as researchers in training and add components as we go through the exercises.

Figure 2.3 A Summative Evaluation for This Cycle

Progress Chart:

Name:

Date:

You may rate yourself on the system you devise. Explain it here and continue the rating process for each exercise.

Here is my assessment of my own work to date:

Exercise: (Name the activity)

Observation of a still life scene:

■ Constructing a Reflective Portfolio

Individuals build portfolios throughout the semester of their work and self-evaluation of their work. This allows for the opportunity to evaluate externally and internally, a requirement for the qualitative researcher. They may add and or remove items as they see fit, and if they need to do more than one observation of a still life or a person, they are free to do so. The contents of the qualitative research portfolio look something like this:

1. Samples of narrative descriptive observations

2. A log of interview activities

3. Samples of transcripts

4. Video and photographic samples

5. Models developed from data

6. Found data poems from transcripts and documents

7. Researcher reflective journal entries

8. Ethical issues in field work as encountered or wondered about

9. (Optional) A website designed to situate all these components and tell your story

10. Reflections on your role as a researcher

11. Description of your theoretical framework

12. A reading list of your favorite books and articles that helped you in becoming a qualitative researcher, including books in various fields as they influenced you

■ Portfolios and Assessment

Portfolios have long been a key method for presenting what a student has learned. It is a multifaceted and complex product. It may have a theme and surely will be judged against a set of criteria usually evidenced in a given rubric. It is similar to the artist's portfolio.

A rubric shows the viewer of the portfolio levels of performance. Constructing portfolios and assessing them takes time, effort, and dedication to the task. Many practitioners define a portfolio as a historical record of student work. It is more than a collection of papers in a folder. It is evidence of a student's work over time and may include accomplishments, capability records, a history of a person's development, and critiques of one's work by both student and teacher. Of course, the items included can be many, but whatever is included should be an authentic measure of what the student has learned. Usually, the tasks that the student performs show evidence of learning and may fall into the following major areas:

1. The tasks performed were done in multiple ways and for a variety of purposes over time.

2. The tasks provide evidence of learning and growth and sample a wide spectrum of cognitive tasks.

3. The tasks show evidence of work at many levels of understanding.

4. The tasks are tailored to the individual learner and offer opportunities for the learner to show what is known.

■ Types of Portfolios

There is no one sacred model or type of portfolio. Depending on the discipline of study (e.g., reading, math, physical education, art, music), the portfolio construction varies. However, in looking over the body of literature on portfolios, there seem to be at least three categories of type of portfolios: the working portfolio, the record-keeping portfolio, and the showcase portfolio. Students invariably construct a combination of all three of these categories for their personal portfolio. See the following:

The Working Portfolio

This type of portfolio is mostly the work of the student, on a daily basis, that gives evidence of ongoing learning in one or more areas of study. Teachers and students freely comment on all aspects of the work. The samples for the portfolio are most often selected by the student, described fully, and critiqued by the student. However, this type of portfolio offers the learner the opportunity to be self-aware and more articulate about learning and his or her own growth process. Many schools begin portfolios at the elementary level and carry through to high school. By the time students come to the qualitative research course, they already have had one or more experiences with the working portfolio.

The Record-Keeping Portfolio

This type of portfolio may be used along with or even integrated into the working portfolio or the showcase portfolio. As the name implies, it is a history of records. It may contain samples of evaluative remarks, assessments, and so on. It is also devised and monitored by the learner with input from the class members and from me.

The Showcase Portfolio

This is the most known and used type of portfolio assessment. Here, the learner constructs a showcase of samples that best describes the learner's progress to date in a given area or multiple areas. It is something like the portfolio a photographer or artist might put together. Usually, this includes completed works that are excellent or outstanding. It is meant to be the record of the student's best work. Thus, we see at least three types of portfolios that provide a record of authentic tasks and learning. Many states have encouraged the use of the showcase portfolio, and in fact, some use the portfolio in electronic format. I offer students the option of doing the portfolio electronically or in hard copy. Most select the electronic portfolio. Now, students may even have their own webpages to upload all their work for restricted viewing within the class and for me. They are actively selecting what to put in the portfolio and webpage and what to leave out or change periodically. It is a dynamic and immediate activity.

■ How the Electronic Portfolio Works

In past decades, with the growth of technology and computers in the classroom, electronic portfolios are a valuable method of assessment. Electronic portfolios allow for easier storage and retrieval of information and also allow for easier inclusion of parental input and feedback. Portfolios are kept on CD-ROM, on websites, or on a thumb drive. A big bonus with the use of the electronic portfolio is the ability to store material that a traditional notebook portfolio cannot. For example, songs, poetry, performances, music, and dramatic readings are more easily stored in digital form, and they capture the activity visually and more accurately. Furthermore, with an electronic portfolio, a new dimension may be introduced: interactivity. Students also may create podcasts or short videos for their websites or for featuring in class. With the wonderful software available for electronic portfolios, students can be more creative and use digital means to verify and adjust their portfolio contents. Most students use a Mac or a PC and update their software regularly. It is always a good idea to invest in good equipment for your work as a researcher. Be aware that technology is advancing so quickly that from the time you enter your doctoral program to the time you finish, your digital voice recorder, iPod, and computer may be outdated. Be sure to get the upgrades you need to be the best researcher you can be. Many ask, Why use an electronic portfolio? The benefits are obvious.

1. Work can be stored digitally more efficiently, allowing for more student options.

2. The display of best examples may be represented more elegantly and more often with the flexibility of the digital format.

Students love it because it allows them to edit, cut, paste, and play back what they have entered. Thus, the prognosis is very good for portfolio assessment. Because most students do their own dissertations on a portable notebook computer anyway, it is no trouble to continually work on your portfolio to showcase your progress in the qualitative research course. Students also remarked that they can do more with field notes on the computer, take notes during interviews, and send reminders to themselves, as well as track other points of data entry, such as references and interview transcripts. In later parts of this text, software for analyzing interview and observational data will be discussed.

Although teachers and students get into the computer age and all that it requires, one can imagine that the transition to electronic portfolios is gradual and not necessarily easy. Yet, the benefits of the electronic portfolio can surely be persuasive as we move to the electronic record. If we ask why we should use an electronic format, many benefits may be listed. For example, consider the following:

- Electronic portfolios foster engaged learning, active learning, and student ownership of ideas.
- Electronic portfolios are repositories of feedback in a medium familiar to many of today's students. This is similar to artistic work as artists have always received feedback on their work and immediately redirect. In dance, for example, notes are given to the performers after each rehearsal and performance by the choreographer.
- Electronic portfolios are the basis for students' discussion of their own progress and a record of their reflections on what they have learned.
- Electronic portfolios are easily accessible, portable, and able to store vast amounts of data and information and remain effective and efficient.

■ Portfolio Contents

The creative activity of constructing a portfolio rests on the learner. Any system may be used. For example, some portfolios are displayed in binders, boxes, display cases, or any combination of the above. Currently, electronic portfolios are on CD-ROMs, thumb drives, podcasts, websites, and can be uploaded at the discretion of the researcher on various websites such as YouTube. Whatever method of display is used, reason dictates that it should be manageable, accessible, and portable. Samples are created, selected, and evaluated by the learner. The contents of the portfolio may include the following:

1. Works in progress, such as writing samples in various drafts and revisions that show evidence of learning. Feedback from the class or the instructor can be included here, as well as the learner's responses to the feedback on the record. This allows learners to revisit this section regularly and recreate portions of it. Thus, the idea that research is a work in progress is punctuated once again.

2. Outstanding products such as poetry, photographs, artwork, descriptions of activities, audiotapes, videotapes, interview transcripts, and field notes. Most important, it should include the researcher reflective journal.

3. Evaluative comments by the student, teacher, and class members. Now, we have the ability to restrict access to anything posted on a website, and obviously, while learning to be a researcher, ethical standards to protect participants are in place.

Remember that the portfolio is at the end of the process chain. Prior to the portfolio development, all of these exercises and opportunities in the field were planned and executed. The researcher reflective journal is alive and changing throughout and is a key component of the portfolio. The portfolio is a perfect spot for the written record of events and what they mean. Portfolios widen the repertoire of reflective assessment strategies and provide solid evidence that students can do this work. It is testament to what the learner has accomplished and learned. Thus, portfolios are a critical element in the qualitative research process. It is extremely valuable throughout the entire dissertation or other research project. To create and shape an individual portfolio is, in itself, a work of art for many. This also affords the opportunity to review and monitor the examples of any work on a regular and sustained basis.

Using what we learned from the still life description, we move on to the next level of complexity: description of a setting.

Exercise 2.2

Physical Description of This Setting

Purpose: To observe one section of the room in which we are seated.

Problem: To see this section of the room.

Time: 20 minutes.

Activity: Select a section of this room that is immediately across from where you are seated. Describe this section of the room in detail. Then, add your reflections on this in your researcher reflective journal.

Evaluation: Continue with self-evaluation and overall evaluation for your journal and portfolio.

Discussion:

    1. How did you approach this exercise?

    2. How is this exercise like the previous exercise? Unlike?

    3. What was the most difficult part of this activity?

Rationale: The movement from a still life description to the setting is a step in complexity. Now, the learner is faced with a number of additional possibilities in how to approach this assignment and how to execute it. Beginning the journey to become a better observer includes this challenge, which is to describe a large physical area. Some individuals add a floor plan and rough sketch of the setting only to return later and measure the floor, the room, the furniture, and so on. Others may take a photo of the setting. Later, the learner tries to capture in narrative form what was captured in the drawing or the photograph.

Exercise 2.3   ∎

Observation in the Home or Workplace   ∎

Purpose: To continue developing observation skill with a setting. Observe an area in your home or workplace.

Problem: To see your own space as you never have before.

Time: 30 minutes.

Activity: Select an area to observe and describe part of your living or working space. Set aside 30 minutes of quiet time to describe a portion of your room or office. Set reasonable goals of description. For example, select one half of the room to describe or one section of the office. Again, after taking notes, type them in field note format and return to class with these data. After discussing these in class, be sure to write in your journal.

Discussion: In class, we will share our descriptions and frame our discussion around the following questions:

1. How did you approach this description of a setting?
2. How did this differ from the previous exercise in class?
3. What was the most difficult part of this exercise for you?

Evaluation: Continue with self-evaluation and overall evaluation for your journal and your research portfolio.

Rationale: By this time, individuals have completed two observation exercises and now move to another level. The learner has a wider band of options to reflect and act upon. Will I select home or workplace to describe? Once selected, on which area of the site will I focus? With the luxury of 30 minutes of observation time, how do I use this time to my best advantage? All are encouraged to think about how this exercise goes beyond both the description of a still life arrangement and a description of the classroom for writing about in the journal.

Exercise 2.4 ∎

Description of a Familiar Person or a Stranger ∎

Purpose: To describe a person sitting across from you, either one whom you know or a stranger.

Problem: To see someone as never before.

Time: 15 minutes. Stop, read to a partner or to the group, get feedback, and repeat again. See if your description changes after the feedback given.

Activity: Select a person to describe physically. Arrange your 15 minutes to your best advantage. Remember to use descriptive behavioral terms, and work for accuracy.

Discussion:

1. What can you identify as major differences in observation of a still life, a setting, and a person?

2. How did you approach this exercise?

3. What was difficult for you in this exercise, and what do you want to do about it?

4. Is there anything about describing a person that relates to your role as the researcher?

Evaluation: Continue with self-evaluation and overall evaluation for your journal and portfolio.

Rationale: The shift from describing objects and settings to describing people is major. For one thing, people are animated. Previous to this, learners had an inanimate, static environment to describe. Now, learners must move to other complicating factors in their observation practice, factors of activity, movement, and life. This exercise is designed to prepare them for observations of people in their future research projects or in the classroom if they do case studies as teacher education projects. I am including an example (p. 54) for your careful study. I mentioned earlier that teacher-educators and preservice teachers often take to these exercises. Here is an example where a middle school teacher went further with the observation of a person in his own classroom. He used this technique to develop his own critical case study. Inevitably, students tell me that this is the most difficult observation exercise because people are always moving, changing, and expressing themselves. Take a look at the critical case of a middle school teacher observing a student and how powerful observation becomes in this case.

■ The Critical Case Use of Observation of a Student: Lessons Learned for the Professional Development of the Educator in Training (Edited)

By Patricia Williams-Boyd (2005)

Professor, Eastern Michigan University

One of the most critical characteristics of effective teaching, particularly in the middle grades, is a deep, abiding, and constantly challenged understanding of young adolescents as complex, unique, and ever changing human beings. Upon the robust knowledge of who students are, how they think, and why they act as they do is built a developmentally and socioculturally responsive pedagogy. This paves the way for teachers to meet the needs of and to set high expectations for all students. Case study research can be a form of action and revelatory research in which the educator-researcher contextually studies a student in greater depth and breadth. The teacher-researcher examines the student both within the community of the classroom as well as within the larger culture of the family. The single case study design and, in particular, the critical case format contribute not only to knowledge and theory building but also uncover areas of weakness, assumption, and limited understanding in the engaged, often hectic, and active environment of the classroom. The case study is a legitimate form of social inquiry because the study is instrumental in furthering the teacher-researcher's understanding of a given problem, concept, issue, or behavior in students. It also represents a contemporary phenomenon—namely, students—in a real-world context—the classroom— where the boundaries between the home, the community, and the school are blurred, given the effects they have on students. Often, however, because of the number of students, the demands of a standards-driven system, and a lack of formal training, teachers see the classroom sphere as existing alone rather than as multiple, overlapping spheres.

When given the charge of conducting a critical case study, my graduate students make some very studied decisions: whom to study, for how long, when during the day, how to ask questions that elicit the kind and depth of information they need but which leave their student in a zone of comfort, how trust can be established so disclosure is not as painful as it could be, how the case study can be written to ethically and powerfully capture the angst and sometimes trauma of the young adolescent world. The master's degree program in middle-level education at Eastern Michigan University includes a course in the theory and practice of middle grades policy, philosophy, program, and practice. Because that course is the first of the cohort classes for middle grades majors, it relies heavily on developing the student's deliberateness both in technique and in perspective. In other words, it asks them to think metacognitively by examining themselves and their students. It asks, Why do you do what you do? Who are the students with whom you work? What kind of environment do you construct for their learning? What variety of instructional techniques is effective for all your students? Why do you make the assumptions about your students that you do? And most important, are there students with whom you feel you have been unsuccessful, students whom you felt you did not reach, much less teach? And did these students, across time, share any common characteristics?

Initially, the most difficult decision the teacher-researcher must make is the selection of one student to study. Separating a piece (a student) from the greater whole (the classroom) is usually based on the teacher-researcher's notion that the selected student is unique or that the student has a distinct identity within the classroom. When the critical case is chosen from the group, we frame the understanding of *group* in sociological terms: a collection of people who interact, who identify with each other, and who share expectations about each other's behavior . . . even if they may not share group membership, in this case, referring to kids who may be left out of peer groups for various reasons. Because the critical case uses the individual student as the focus of the study, it is often called a microethnography.

Critical Case Study ■

Purpose: To conduct a critical case study. This will lead learners to the point at which they can construct their own meaning and answers to these questions. The study tends to be illustrative and is conducted inductively. Induction helps with understanding the qualitative research process.

Problem: To examine the classroom as a constellation of individuals rather than as a group of young adolescents, to confront myself and challenge assumptions, and to develop a sense of self as a teacher-researcher and as an action researcher who can identify and solve problems.

Time: Gather data in the field for 6 to 7 weeks. Leave the field, and begin the analysis and interpretation of data.

Activity: Select a student who causes you to ask some critical questions. Gather data through personal interviews, and triangulate both the source and interpretation of data. Present your study in five to seven pages with a minimum of seven bibliographic sources. Use respected journals in the field. Use APA style. Your paper must have a title page and must have a creative title. The body of the paper should follow the research design: study question, propositions, the student (with care and protection of the student's identity and with written permission from the student, the family, and the building principal), research interpretation, theory, and conclusions. Address key issues or general concepts. Organize your paper in the following fashion:

- Start with the presentation of the case. Describe the boundaries of that case.
- Next, present the research interpretation of the case.
- Then, discuss the extension of the study or the ways in which the study has influenced you as a professional.
- After the paper's conclusion, offer several discussion questions that will help the reader focus his or her own perceptions of the study's key issues.
- Finally, present your bibliography and suggested readings if they are significantly different from the works cited.

Discussion:

1. How did you select the student you studied? What criteria did you use?
2. What kinds of questions did you ask that offered you the depth of information you sought?
3. How did you establish trust? How did your relationship with the student change across time?
4. What did you notice about yourself, both personally and professionally, as you moved through the process?
5. What challenges did you encounter?
6. How did this study affect your professional development and relationships with students?

Evaluation: The case study will become part of the student's master's in middle-level education portfolio. As well, both the professor and the teacher-researcher will use the assessment tool described in the syllabus to critique the study. For example:

1. Meets requirements	_____	25 points
2. Format	_____	25 points
3. Textual presentation	_____	25 points
4. Research interpretation	_____	25 points

Rationale: This tends to be a pivotal experience for master's students. It provides the opportunity for them to experience the power of their own research and analysis to more fundamentally understand who their students are. Consequently, the learning activities that they construct will look different. They see themselves as advocates for their students, for they challenge themselves to develop a more responsive classroom. And, they talk about themselves as teacher-researchers, as action researchers who are agents for change.

In the analysis of what is required in both the final presentation as well as in the conduct of the study, my students and I engage in a collaborative discussion of the skills requisite to an effective case study. Together, we decide on the following:

- The researcher should be able to ask good questions—questions that are open-ended—encourage the student to disclose an often sealed-off area of his or her experience, and move the student incrementally from his or her zone of comfort. The researcher should then be able to interpret those answers.

- The researcher should be a good listener and should not be trapped by preconceptions. The thoughtful researcher listens for how things are said, watches for body language cues, and often listens for what is not said as well as what is said.

- The researcher should be adaptive and flexible to the extent that newly encountered situations can be seen as opportunities rather than threats or deficits.

- The researcher must have a firm grasp of the issues being studied.

- Perhaps even more important, the researcher must be willing to deal simultaneously with multiple ideas and possibilities and must be comfortable with ambiguity. The researcher's willingness to research the literature in an ongoing fashion helps him or her to better understand the unexpected twists and turns of interview data and better positions the researcher for the next interview session.

- The researcher should be aware of preconceived notions, including those derived from theory. Therefore, the researcher should be sensitive and responsive to contradictory evidence. Look for what does not make sense.

- The researcher must be respectful of and sensitive to the total person being studied.

■ The Qualities of Exemplary Case Studies

The Case Study Must Be of Interest to the Researcher

- Underlying issues you suspect may be present are globally important to your field in either theoretical or practical terms. The individual case must be unusual, not typical of the average young adolescent, and it must be of general public interest.

- The case may be revelatory in that it reflects a real-life situation that other social scientists may not have been able to study in the past. It will be revelatory for you as the researcher in training.

- Consider in advance what contribution to the theoretical as well as to your professional understanding of young adolescents your study will make.

*The Case Study Must Be Complete
and the Boundaries Defined*

- Give explicit attention to the boundaries of the case or the distinction between the phenomenon being studied and its multiple contexts. Describe what the boundaries of the case include.

- The collection of evidence must be thorough, relevant, and sufficient, and it must be triangulated.
- Judiciously and effectively present the most compelling evidence so the reader can reach an independent judgment regarding the merits of the analysis.

The Case Study Must Be Composed in an Engaging Manner

- The reader must be able to see the person and the environment as described through the use of rich and descriptive language.
- Although many design models are effective, the following is a friendly way to introduce practitioners to the dynamic world of qualitative analysis.

The Research Design

The Case Study Should Answer the Following Questions:

- *Who:* Explain who the individual is and what the immediate setting looks like.
- *Why:* Describe why you chose that particular student, why you are doing the study, and what if any changes you propose making at the conclusion of the study.
- *How:* Discuss how and where you are going to conduct the study, what questions you will use, and how you are going to develop some assumptions that you will interpret.
- *Where:* Describe the sociocultural context of the classroom, the school, the family, and the immediate community.

Making Sense of Data

- Engage in research.
- Interpret the field research through literature.
- Develop your own theories, and test them against the literature.
- In explanatory cases, examine the facets of your argument.
- In exploratory cases, debate the value of the work.

Criteria for Interpreting Findings

- Match patterns; look for points of tension in the data rather than for a single, consistently explanatory conclusion; look for emerging patterns or themes; triangulate data sources; and find multiple analysts such as peers, other teachers, and parents.

- Develop analytic statements based on physical and cultural artifacts as well as on fieldwork.
- Draw conclusions.
- Ask what implications this new knowledge has on your own practice; what professional as well as personal changes did you experience?

The following is an exemplar of the critical case study based on observation of a student and following Professor Williams-Boyd's (2005) format.

People make sense of their world as filtered through the lens of their own lived experiences. Teachers make sense of the behavior of their students through their own beliefs, expectations, values, dispositions, and assumptions, in addition to their experiences. The use of critical case studies not only informs classroom practice and pushes teachers' professional growth, but it calls them to action and advocacy on behalf of their students. During the 7 years I have used critical case studies with graduate students, as they begin to unlayer their own complex students, I have seen teachers shift from being somewhat passive consumers of higher education to becoming engaged in active agency.

Outstanding teacher-researcher Kimberly Toloday (2002) writes in the opening of her critical case study:

My greatest regret as a middle school teacher was not truly understanding middle grades students and not getting to know them as individuals. I taught all seventh grade core subjects except social studies, and I was overwhelmed with adolescent attitudes and seven different daily preps.

Teacher-researcher Matthew Admiraal (2000) thoughtfully and passionately concludes in his case study:

Working with Phil has been a privilege. He has made me much more aware and understanding of the difficulties faced on a daily basis by a person with ADHD and a learning disability. Through this experience (the critical case study) I have learned the need to identify and to adopt interventions to better help Phil and other students with ADHD. I desire for him to succeed academically and socially and hope that with proper support he will overcome the double challenge of ADHD and a learning disability.

■ Missing Dad

By Thomas Beltman (2002)
Middle school teacher

Presentation of the Case (Edited): An Example

Teachers' days are so often busy, and everything is urgent. Still, there are sometimes blanks in the day, utterly empty moments. At those times, I am neither preparing nor finishing an activity. I am just there. These moments are like dead air on the radio—they come infrequently and unexpectedly. One comes today, between first and second hour when there is no one in the room for just the briefest moment. I briefly pause to feel the shape of suspended time, expectant.

Now, they are coming, and the bubble pops. Soon, the room is half full, 12 of 27 eighth-grade students. Brooke teases A. J., who steals her pencil but gives it back. One student comes in and drops his books from too high onto the table so they bang, then gives them a shove to see if they will slide all the way down to hit the wall. His seat is there, and today he has to pick up his books from the floor. "Why do you do that?" someone asks. "I don't know. 'Cause it's fun." And, I'm thinking, "to see if anyone will notice."

I've previously seen Kenny, an African American in eighth grade, do this and some other things to gain attention: drum on the table, talk a little louder than the other kids, or try and insinuate himself into their conversations. He disrupts class by talking out, or through, sub-conversations. He cannot be stopped with "The Look"; it takes a conversation. He is not popular, though he is not shy; most students know him. He is sometimes laughed at because he has a slight speech impediment. He is overweight and a bit clumsy. He carries his head up, and his back is straight. He is proud. His ideas for what he will make usually exceed his ability. He appears bold, but his confidence is thin. He is usually cheerful and jokes with me. Kenny is aware of what is going on around him and offers good advice to me and to his peers. For example, when I had trouble with the tape recorder one day, he said, "I think you should just take notes. You can't rely on technology these days." There is something about him, a hint of waiting in his eyes and demeanor. He's looking for something and isn't sure if he finds it in me, or if I can be trusted. This boy puzzles and intrigues me, and I would like to know more about him. I chose to interview him because he is often in the building early, and his disruptive and defiant side may point to deeper issues. What I discovered was a very complex young man, responsible and lonely, cheerful and hurting.

I had him in my class last year, and he didn't pass. In his other classes, he was pulling Bs and Cs, and an occasional D or A. He reported that he did very well on the standardized tests he took in sixth and seventh grades, and his sixth-grade teacher said he was smart. He said he doesn't do as well as he could because he finds it hard to stay focused all the time, because teachers do so much yelling. In my class, completing the assigned projects, keeping records of the work, and handing them in at the end of the project ensure success. He was among the minority that loses part or all of these records. He is, in this way, self-defeating. I thought then he was just lazy. It turns out I wasn't only wrong; I was being very unfair. Without knowing his story, I had assigned him a burden of fault that he did not deserve. What he needed was support, not judgment. I couldn't have known this without getting to know him better.

Unfortunately, he wasn't very willing to talk about himself. I had to approach him obliquely, as though with a sideways glance, couching my questions as inquiries into what he perceived other kids thought and felt. Since I was asking him to render opinions about the thoughts of someone else, a topic he couldn't know with certainty, I assumed he would be projecting his own experience into their circumstances.

Kenny was willing to tell me a little about his family—he lives with his mother, and he has two brothers and a sister living with his dad in Maine. His mother and father never married, and his father moved away when Kenny was very young. He gets some gifts from his father on his birthday, and talks to him on the phone a few times a year. His mother has a sister whom they see frequently, but her kids are much older than Kenny. His mother has worked different jobs throughout his life, and he has spent much of his early nonschool time in latchkey programs at his elementary school. He does not view this as bad. He thinks his mother was doing what she had to in order for them to have a roof and clothes. Though he missed her sometimes, he had lots of playmates and was busy.

(Continued)

(Continued)

Now, he comes home to an empty house and reports that he does his homework right away for about an hour. Then, he does some TV watching and makes dinner. In the summer, he mows the lawn, and in the winter months, he keeps the walks and drive clear of ice and snow. Sometimes, he does his own laundry. His mother often does not come home until late, because now she has a job at her boyfriend's company. She has laid down some strict rules about what he can do— no friends over if Mom isn't home, and he must tell her the day before if he is going to a friend's house, or call her at work. If he can't reach her, he can't go. He hates it that she wants to control his every move (in his view). He would prefer to have more freedom, like some of his friends, but he also doesn't like the trouble they get into when they are exercising their freedom.

Kenny feels that most people his age don't know what they think or who they are. I take that to mean he doesn't know who he is. He also feels that most kids his age, with some exceptions, don't really fit in well with other kids. He thinks this is true even if they look like they have lots of friends. He thinks maybe girls are different, because they are "more emotional about stuff so they make better friendships." I may be wrong, but I feel Kenny is talking about himself again. This feeling is based in part on my observations of him in my class.

Because my class is mostly lab time, I have the opportunity to observe peer relationships in their near-natural state. The room is large, noisy, and full of activity. After a few weeks, I can sometimes fade out of their radar and watch them interact. I did this with Kenny and noted that the other kids didn't seek his space; he had to go to theirs. He isn't a follower in the strictest sense, but neither does he go his own way. He said, "Kids tend to be different people with whoever they're with. You know, like when I've got [sic] in trouble, it's the kids I'm with. I'm not always like that. I don't know. It's weird. I'm not saying it right."

Research Interpretation of Cases (Edited Example)

The first thing I noticed about Kenny this year was his awkwardness around his peers. It wasn't that they didn't like him; it was that he didn't know how to relate to them easily. He lacked the social skills they had, and they capitalized on his clumsy tenders of friendship by pushing him to the outside edge of the circle of intimacy. After learning that he had no strong male role models in his life, I wondered if there might be a connection. Research suggests that children who become father-absent before the age of 5 suffer "debilitating intrapsychological and interpersonal difficulties" throughout their lives. This is more pronounced for boys. Research also shows that father absence could seriously affect the sex role development of boys.

Writers have suggested that the role of the father in a boy's overall development is very important. Not only does the father provide a role model for actions and behavior, but he also provides a model for a boy's emotional life. For a boy with an absent father, the loss is felt both in finding a place in the world as a man, and in understanding himself and others emotionally. The quality of peer interactions seems to be affected in some way for boys such as Kenny.

Kenny has considerable responsibility for himself at home after school and strict parental parameters on what he may do. This pattern is supported by research which shows that contrary to the general assumption that single parents exercise less control over their children, there is no significant difference between how two-parent families and single-parent families exercise control. In fact, it is often the case that single parents are more likely to set strict curfews and domestic activity guidelines.

In any case, Kenny's mother must be doing something right, because he is beating the odds. He is not delinquent, and he is not failing school. Surprisingly, 46% of father-absent, African American adolescents repeat a grade. In addition, they experience considerably more behavior problems than do father-present adolescents, particularly in terms of running away from home, skipping classes, being suspended from school, and getting in trouble with the law. It is possible to overcome these grim statistics with adequate supervision. It may seem impossible for a single parent to provide enough guidance for her son. In fact, conventional wisdom is that single-parent households would have less parental supervision than

(Continued)

(Continued)

their counterparts, but research has shown otherwise. Another significant area of adolescent development that may be at risk for the parent-absent child is in the area of identity formation and locating the self in a social context. Basically, kids need their parents while growing up.

Conclusions About the Cases

Kenny got an A in my class this time around, in stark contrast to the E of last year. I will probably see him again for the other eighth-grade rotation class I teach. He will find the class grading structure changed because, as a result of this interview, I have made some changes. Instead of handing out a packet and collecting it at the end, I give out the work in pieces and collect it as we go. The old way merely punished students like Kenny, who were coping with a lot, too soon in their lives. I undertook this interview with Kenny in order to better understand his circumstances and the factors that influence his personality development. I found a young man who is an amalgam of complex and sometimes conflicting influences. He harbors contradictions and secrets, no less jarring or cherished than my own.

Discussion Questions

1. How can you structure your coursework so that students with poor organizational skills are not unduly penalized for forgetting or losing work?
2. In what ways could your perceptions about single-parent families color your relationships with students?
3. What can schools do to help fill the absent father void, particularly for African American boys and what can teachers do to help students in this group?

Here then we see some of what Thomas thought about as a teacher of a father-absent male student. He left us with some good questions to reflect upon. Common to most critical case study research is the concluding self-critique. Teacher-researcher Jodi Tye (2002) sums up what graduate students who have engaged in critical case work contend when she says the following:

> Speaking with Sarah over the last couple of months has continued to help me grow as an educator and a person. Although at times it is difficult to have a child in my classroom with a disability or disorder, I feel more often I learn from those children. I learn not only about dealing with children or being a teacher but about being a human and dealing with people.
>
> At the beginning of the case study assignment, I ask my graduate students to choose one of their students with whom they will study. I wait a moment for the quizzical looks to turn to understanding. "Yes, one of the required pieces of this assignment is your approach to your chosen student. You will undoubtedly find it helpful to ask the selected student if he or she is willing to spend time with you in order to teach you more about young adolescents." Inevitably, my graduate students come back the next week and share that, to their amazement, the students they chose felt honored. And they share that when they asked the students if they would feel comfortable with a reversal in roles, the students seriously contemplated what that could mean, and without exception, the students were willing to help their teachers better grasp a world about which they thought they knew. The teachers soon realize that it is a world that will open up new understandings that will change their teaching careers. The critical case study is the shared space in which the teacher-researcher becomes a quiet, listening, observant participant, and the student reveals new perspectives. Critical case studies may be the pivotal experience that moves teachers from passive instruction to active agency, from assumption to informed and inspired understanding, and from the desire to know the links between teaching and learning to discovering the links between cultures of experience in the lives of their students—an understanding that informs teaching and learning.

Thus, from Pat's work as a teacher-educator of middle school teachers in service, we see some fine examples of how observing one student with care and precision can assist in the development of the practice of teaching. This alone shows the many practical applications of observations of a person for teacher-educators. Next, we go on to description in a public place, the art museum.

Exercise 2.5 ■

Observing at an Art Museum *or at a Movie* ■

Purpose: To observe and describe two areas of an art museum, such as the lobby or gift shop, or an area of the museum where you find two paintings, sculptures, or any works of art of interest to you. Alternately, if you have no access to an art museum, then a movie will be a good substitute. Currently, I am fortunate to be able to have a field trip with my students to the Salvador Dali Museum nearby.

Problem: To see the art museum as a place of movement, activity, silence, and so on. Watch for interactions of people with artworks. See this place as you never have before. If you go to the movie house, apportion your time to describe the movie house, the movement of the patrons, and the movie itself. Recently, for example, students observed the movie *The Blind Side*, observed and described the patrons in class and the movie house along with the setting of the movie, and selected one character from the movie to describe. If selecting the movie, obviously stay for the entire film.

Time: 15 minutes for each of the three areas, 45 minutes total. For the movie, stay for the entire film and do the best you can once you select the one character to describe.

Activity: Take time to select and focus on these public spaces. If you need to do so, that is, if someone asks you what you are doing, introduce yourself and explain that you are doing this to become a better observer. Assure them of confidentiality, that no harm will come to them, and that you are doing this for a class. Most people will be delighted that you are there. In fact, you may have to stop them from speaking to you so that you may go on with your observation. After 45 minutes of observation and field note taking, return to the group—if you have the benefit of a class, for example—to discuss this experience. If on your own, take a few minutes to write down your thoughts about this exercise. Of the two exercises, students love the movie exercise better, and if they select the art museum, they always say they wish to return and redo the entire exercise. They also remark that they are forever changed, and at every subsequent visit to the museum, this exercise has changed how they view art and how they view the context in a public space.

Discussion:

1. Identify what you were thinking as you selected your areas.
2. How did this differ from previous observations?
3. What was most difficult for you in this activity?

Evaluation: Continue with self-evaluation and overall evaluation, and write in your researcher reflective journal about this exercise.

Rationale: Learners move to yet another level when they begin to realize what it is like to observe and describe in a public space. They begin to realize what gaining access and entry to a social setting feels like, although admittedly, they are in a safe environment at this point, and one that is somewhat familiar to them. In addition, learners have firsthand experience in articulating their goals of observation. They are working on multiple levels of observation and description. I like to suggest finding a quiet place to work afterward to reshape the field notes, review notes to the self, and render the description in standard sentences and grammar. This will help later in studies when an individual is trying to make sense of mounds of data.

Exercise 2.6   ■

Observing an Animal at Home, the Zoo, or a Pet Shop   ■

Purpose: To describe an animal at home or at the zoo.

Problem: To see and recognize the complexity of the animal's movement.

Activity: Go to the zoo or a pet shop, or observe an animal in your household. Physically describe the animal—how the animal eats, moves, and shifts position and attention. Can you make a list of five adjectives that describe this animal? Do you see inner qualities as well as outer qualities? Isolate your description to include limbs, eyes, head, and mannerisms. Find two traits of the animal that you also find in human beings. Construct a metaphor for this animal. Use the following as examples:

This cat is a princess. (Then, the writer describes physical and behavioral characteristics to convince us that this cat is a princess.)

This gorilla is a prizefighter. (Again, the description of physical and behavioral qualities should convince.)

This dog has become the caretaker of the family. (Continue with physical descriptions, and describe interactions with family members.)

Time: Take as much time as you like with the observation.

Discussion:

1. What did you notice about yourself as an observer as you began this exercise?
2. What was difficult about describing an animal?

Evaluation: Continue with self-evaluation and overall evaluation for your journal and portfolio.

Rationale: Individuals may learn something about the difficulty in observing people by beginning with observing animals. Like human beings, animals are animated, unpredictable, and active. This is a good beginning step in complex description in a complex setting. Learners often remark that, although difficult, this exercise was challenging and enjoyable at the same time.

Exercise 2.7   ■

Drawing to Become a Better Observer: Drawing Upside Down   ■

Purpose: To try some drawing exercises, which are the heart and soul of observation, in this case, an upside-down drawing of a famous painting.

Problem: To see this painting and view it in a new light—upside down.

Time: 45 minutes.

Activity: I use a drawing by Picasso (1881–1973) suggested by Betty Edwards (1999) in her text *The New Drawing on the Right Side of the Brain.* The drawing is a pencil drawing called *Portrait of Igor Stravinsky* done by Picasso in 1922. There are variations of this exercise in virtually every art book on drawing. I amend it here to serve as a strategy for developing the role of the researcher, to awaken one's artistic intelligence, and to serve as a catalyst for good narrative writing. Learners are given a copy of the drawing and asked to turn it upside down. They often mention they simply cannot draw, and some teacher in their educational history made note of that fact.

Instructions:

- Copy the image as you see it upside down. Use your pencil and your eyes to follow the lines, starting anywhere you wish. Find your personal velocity, and start where you decide is best for your concentration level.
- Use a pencil and plain white paper.

(Continued)

(Continued)

- Copy it any size you wish.
- Start drawing at any point you wish; there are no sequences. Some start at the top and work left to right, others start right to left, some start in the middle and carefully add lines on one half of the page, and so on. Some start with an article of clothing, say the jacket sleeve. All the while, attention is focused on the upside-down drawing. Do not turn it right side up until you are finished.
- When complete, turn your drawing right side up.
- Begin discussing the exercise. Students are amazed when they realize that indeed they can draw if they are careful with observation, they have the time, and they concentrate.

Many students are so inspired by this exercise that they actually devise some additional work of this nature and share it with the class at future meetings. The point most often made in the discussion that follows is that there is a strong tendency for some to want to turn the drawing right side up. I might point out here that the silence in class is complete and utterly poignant, so intense is the concentration in this exercise. For qualitative researchers, it is a valuable exercise because it shows clearly how we can switch modes in our thinking. In the field, one is often required to do this at various points in the research process. Just as you are forced to think differently when viewing something upside down, you see things differently. For the qualitative researcher, this awareness is helpful later in the research process, because we are called upon at the oddest moments to see things differently. After doing this, learners realize that in fact they can draw, that they need to practice in order to draw, and that they need to see what is in front of them in order to draw it! Now, you have progressed through seven exercises that will sharpen you as a research instrument. You are developing the habits of observation and reflective journal writing, which will prepare you for the exercises in the next chapter to advance these techniques.

CHAPTER 3

Advancing the Observation and Reflection Habit ■

The difference between the right word and the almost right word is the difference between lightning and a lightning bug.

—Mark Twain (1996)
American Author and Humorist, 1835–1910

Now that you have completed the exercises in Chapter 2, you may be happy to know that you are ready to go into the field in a public space. You have successfully completed a series of observation exercises that took you into various levels of difficulty, starting with the observation of a still life scene. You progressed to describing people and action, then drawing, and now it is time to test your abilities as a research instrument and start a new habit. This first exercise is for developing the habit of nonparticipant observation in a public setting and strengthening your beginning habits of reflection and journal writing.

Exercise 3.1 ■

Nonparticipant Observation Assignment ■

Place: Restaurant, coffee shop, shopping mall, book store, zoo, place of worship, museum, health club, funeral parlor, dog park, beach, skating rink, park, movie theater, library, technology center, or any public setting.

Purpose: To observe, describe, and explain a complex public setting. There should be natural public access to the setting and multiple viewing opportunities for you.

Activity: Nonparticipant observation. Go to this social setting more than once in order to get a sense of the complexity and to maximize what you learn. Go at least three times at different times of the day. If you wish to return at any other time, of course, feel free to do so. Take notes. Make a floor plan. Discover what you are able to hear, see, and learn just by observing. Take photographs if possible, but ask permission first.

What to Look For:

1. The setting: Look around you and describe the entire physical space. Draw a floor plan or take a photo if permitted.
2. The people: Look around you and describe the people in this setting. Focus on one or two of the people. What are they doing in that social space?
3. The action: What are the relationships between people and groups? Try to discover something about the people in the setting.
4. Describe the groups and any common characteristics, such as age, gender, dress codes, speech, activity, and so on.
5. Focus on one person in your viewing area to describe in detail, for example, a waitress,

a caretaker, or a salesperson, depending on your
setting.

6. If you had all the time in the world to do a study
here, what three things would you look for upon
returning to the setting?

7. Be sure to <u>give a title to your report that captures
your study</u>.

8. Be sure to use references from a minimum of seven
texts.

Time: You have 4 weeks to complete this assignment. Be sure
to include a self-evaluation.

Discussion:

1. Of all the exercises so far, how has this one
challenged you?

2. How did you approach this assignment, and what
difficulties did you encounter in the field setting?

3. What would you do differently if you were to return?

Evaluation: Continue self-evaluation. And at this point, you need to
complete a summative evaluation of this observation
cycle for entry in your journal.

Rationale: At this point, the individual has direct experience in
nonparticipant observation, observation of action, and
self-evaluation and has been given the opportunity to
discuss with peers his or her progress to date. These
exercises give the learner a taste of what it will be like in
the field in terms of the ministudy for the class and quite
possibly in terms of his or her own dissertation or future
research activities. Also, there is growth in terms of
thinking like a researcher and developing a role for
oneself. In the context of a course, the ministudy may
be used as a pilot study for a dissertation project if it is
reasonable to do so. The learner also begins a research
reflective journal and starts, in most cases, an electronic
portfolio. No matter what the situation of the individual
reader of this text, throughout the practice of these
exercises, learners must continue reading from a series
of reading lists. Here is an example of a completed
project. It is edited in some portions. It was selected
because the salon caters to both men and women.

■ The Cutie Beauty Salon

By Jason Pepe (2007)
Doctoral student in educational leadership
and policy studies

The Setting

The Cutie Salon is located at 10841 Bloomingdale Avenue in Riverview, Florida. The family-owned business is nestled among several other shops in the Albertsons strip mall. The strip mall, known as Bloomingdale Hills Park, is situated on the busy intersection of Bloomingdale Avenue and Providence Avenue and can be entered from either road. In addition to Albertsons Supermarket, this large strip mall is home to independently owned businesses such as the Cutie Salon as well as franchises such as McDonalds, Wendy's, Subway, 7-Eleven, and others.

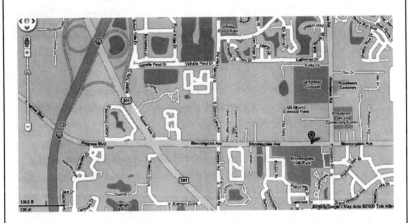

Cutie Salon, 10841 Bloomingdale Avenue, Riverview, FL 33578

Cutie Salon sits on the south side of the strip mall, directly to the west of Albertsons Supermarket. It is sandwiched between two other stores; on the left is Dental Health Group, and on the right is a pizza shop no longer in business. Each business has its unique style of lettering on front side of the overhang. The words *Cutie Salon* are in all red capital letters centered over the front entrance.

Cutie Salon Storefront

The overhang (or covered walkway) is supported by concrete block pillars running the length of the mall, one approximately every 15 to 20 feet. The ceiling of the overhang is lined with aluminum siding along with evenly spaced, recessed lighting. The covered walkway also includes rectangle-shaped signage for each business. Although the shape of these signs is identical, the lettering matches that of the larger sign on the outside of the overhang.

The storefront of Cutie Salon is separated into four glass panels and a door. There are two large glass panels to the left of the door. Adhesive block lettering spells "Hair, Nail, Facial Waxing" in white capital letters. Two red neon signs hang directly above the lettering. The first sign reads "Open" inside a blue circle. The second neon sign reads "Hair" inside an orange rectangle. Smaller white letters cover the door indicating shop hours. It also states "Walk-ins & Appointments Welcome" along with the phone number. The same white adhesive block lettering covers the last glass panel to the right of the door and reads "Hair Cut for Men, Women and Children." Another neon light sits above these letters, spelling "nails" in all capital letters inside a blue rectangle.

(Continued)

(Continued)

Upon entering the Cutie Salon, I immediately notice that the shop is in the shape of a rectangle and divided into two halves. The facility measures 60 feet in length by 20 feet in width. The front two thirds of the store is an open space where most of the business is conducted. The last third is separated into various rooms. A kitchenette, a break room, and a bathroom are located on the right. A massage room and storage room are located on the left. A back door (for employees only) opens to a small parking lot not visible from the front of Bloomingdale Hills Strip Mall.

Unframed mirrors and posters

Leads to bathroom, storage, kitchenette, and employee parking

Extension cords, multiple outlets, and connections

Manicures and Pedicures: The Right Side of Cutie Salon

Customers who want their fingernails or toenails trimmed and polished spend their time on the right side of Cutie Salon. A large wooden shelf sits against the wall in the front of the store. Five or six piles of magazines clutter the top shelf, and four telephone books are stacked on the bottom shelf. A large mirror (approximately 5 feet by 3 feet) with a decorative wooden frame hangs above the wooden magazine shelf. To the left of the large mirror are a series of three unframed mirrors, a poster, one unframed mirror, another poster, and two unframed mirrors. Plastic shelves directly below the unframed mirrors hold a glass vase with fake flowers and various nail tools and beauty products.

Manicure customers sit at one of four separate nail stations. The client faces opposite the nail technician in black leather office chairs with armrests and a mechanism to raise and lower the height of the chair. Polish, nail tools, brushes, chrome containers, powder, lotion, and various other objects scatter the desks at each station. A large pole (similar to those used in hospitals to hang IVs) is perched next to the station, holding an electric instrument used to file nails. The look and sound remind me of a dentist's office. A large white light with a hinge arm and a small electric fan are positioned on the left corner of the station. A mass of wires resembling spaghetti pours over the station, converging at a four-plug outlet on the side of the desk. Extension cords connect to each other in a long series, running the length of both sides of the salon. Although plastic cord covers placed between stations prevent tripping, I wonder if the multiple wiring setup violates fire codes.

Pedicure customers sit in one of three large, maroon recliner chairs. The recliner includes many special features. The armrest lifts up and down to allow easy entry into the comfortable chair from the

Spa pedicure chair with attached foot bath

(Continued)

(Continued)

Recliner chair with leg extension to wash hair

side. Small gray tubs attach to the bottom of the recliner. The tubs fill with warm water for foot soaking. The nail technician sits on a small leather ottoman with a back, where he or she washes, trims, polishes, and paints customers' toes.

Directly to the left of the three spa pedicure chairs are two sinks used to wash customers' hair. The black leather chairs recline into the sink, and there is an extension for the customers' legs during hair washing and rinsing.

Color, Cut, and Blow Dry: The Left Side of Cutie Salon

When a customer visits the Cutie Salon for a haircut, she or he sits in one of five stations on the left-hand side of the shop. Identical hair-styling stations line the left wall with a corresponding pedestal chair. A pressboard cabinet with a cherry veneer covering surrounds a large, rectangular mirror. Three shelves to the right of the mirror hold shampoo, styling gel, powder, and various beauty products. Two recessed circle lights shine above the mirror and above the three shelves. Straight irons, curling irons, combs, and electric trimmers are strewn on the counter beneath the mirror. A hair dryer is holstered inside a hole on the right side of the counter. Two drawers with chrome handles built beneath the counter house extra beauty supplies. Electrical devices plug into a four-plug power adapter under each cabinet. Similar to the right side of the store, a tangled web of wires connect to several extension cords, forming an electrical grid down the left side of the store. A plant decoration sits on top of every cabinet. However, the third cabinet houses a 32-inch flat-screen television tuned to the Turner Broadcast System (TBS). I sit in the customer's chair and find it quite comfortable. The black leather chair with arm-rests sits on a chrome pedestal. The hairstylist uses a chrome foot pedal mechanism on the base of pedestal to raise and lower the customer.

Before a customer sits to have his or her hair cut and styled, he or she usually waits in one of four chairs pressed against the front glass panels of the salon's storefront. These rather uncomfortable chairs are covered in black vinyl with chrome legs and no armrests. Two glass shelf units are positioned to the left of the waiting area. The first shelf is constructed of glass, supported by chrome bars about 4 feet high. Various hair products are neatly stacked on each of the four shelves, including Biolage Matrix shampoo and conditioner, Redken anti-frizz polishing milk, Aquage color-protecting, seal-in treatment, and Matrix

(Continued)

(Continued)

Curl Life hair spray. I am not familiar with any of these hair products, and I believe they are only available for purchase in salons.

The second shelf is a rectangular glass unit with 15 glass boxes (3 boxes wide, 5 boxes high). The shelf unit rests on top of a black wooden base. Chrome bar supports connect the boxes together. When I look inside each box, I find a wide array of hair products unfamiliar to me, including Vavoom smoothing gel, Ice Hair colored styling glue, American Crew styling gel, Matrix Biolage hair spray, and Amplify volumizing system. For a brief moment, I wonder if one of these products is the answer to my thinning crown.

One of the most interesting objects in Cutie Salon is positioned between the two glass shelf units. The careful observer notes a small, red wooden shrine. I kneel on the floor to take a closer look. At the corner of the red box are two silver urn-shaped candleholders deco-

rated with protruding dragons. Inside each holder is a red plastic electric candle embossed with silver dragons and illuminated with yellow lights. Behind the right front candle is a ceramic old man dressed in orange and red robes. His long, grey beard flows down in front of him. He grasps a gold object in one hand and a red staff in the other. His grey head is adorned with a matching skull cap.

The laughing Buddha sits at the right hand of the old man. He rests in the classic Buddha position, hands casually on his knees and his round belly between his legs. Above his happy eyes and around his bald head is a red band that matches his red robes leisurely draped around his shoulders and arms. A cracked, red fan held together with Scotch tape rests in Buddha's left hand, covering his nipple while exposing a gem-adorned right nipple. The empty hollow of his right hand suggests something may be missing.

At the Buddha's feet is a dirt-filled, brown ceramic pot with 20 or more red incense sticks. The ash ends of the burnt incense litter the top of the pot and the inside of the shrine, dusting the Buddha's red ceramic robes. Two clear shot glasses filled with water sit on each side of the incense pot next to the candles. On the tile floor outside the right corner of the

shrine sit two green pears and a mango placed in a cream-colored, plastic oval plate. A single, tattered stalk of lucky bamboo extends from a white ceramic vase with blue painted flowers to the left of the fruit offering. A dirty tangle of orange and white extension cords are bunched between the wall and the shrine. In the very back of the shrine, tucked neatly behind the two men, are two lottery tickets, numbers darkened for a hopeful win.

People

The same five employees, three women and two men, work at the Cutie Salon during both my observations. All five employees are of Asian descent and speak English as a second language. During my initial observation, I ask one of the employees at the front counter for permission to observe in the salon. She points to Employee 1, a nail technician, and instructs me to ask him because he's the owner. Employee 1 stands approximately 5 feet 1 inch, is roughly 125 pounds, and is approximately 50 years old. He has short, black hair peppered with grey along the sides. A thick, groomed mustache covers his upper lip, matching his thick eyebrows, which come to a point on the bridge of his flat, broad nose. Employee 1 has a pitted facial complexion and straight teeth. It is very difficult to understand his broken English; however, we manage to

(Continued)

(Continued)

Employee 1

have a conversation about this assignment, and he graciously agrees to my request. Most of Employee 1's time is spent sitting in the first nail station, doing manicures for customers; however, he does cut and wash hair.

Employee 2 is an Asian female standing approximately 4 feet 11 inches, weighing roughly 130 pounds, in her late 40s or very early 50s. She has straight, shoulder-length black hair with three or four streaks of light brown and a part down the right side of her head. Her eyebrows form a perfect arch over her dark brown eyes and look painted on her forehead. Employee 2 has straight teeth with a slight overbite and a fairly smooth complexion, and she wears more makeup than the other two female employees. She is observed cutting and styling both women's and men's hair, as well as trimming and polishing fingernails.

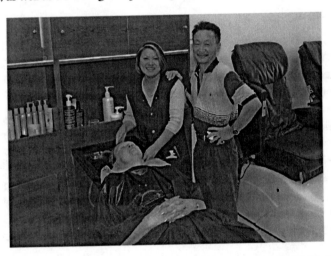

Employees 2 and 3

Employee 3 is an Asian male in his early 50s, stands 5 feet 2 inches, and weighs approximately 120 pounds. Crow's-feet gather around his dark eyes, especially when he smiles. He keeps his black hair spiked longer on the top and trimmed short over his ears. A few grey hairs are scattered amongst his mostly straight black hair. A very thin, neatly groomed moustache traces his upper lip. A few freckles and spots dot his face, especially on his upper forehead above his small, half-pointy eyebrows. Employee 3 is observed cutting men's and children's hair.

Employee 4 is an Asian female in her mid to late 40s, approximately 4 feet 11 inches, and 140 pounds. Her smooth, round face and light complexion are in stark contrast to her jet-black, shoulder-length hair. Employee 4 has dark eyes, very even eyebrows, and a broad nose. She is observed performing manicures, pedicures, and haircuts on women, men, and children.

Employee 5 is observed the least amount of time. She is an Asian female in her mid-30s, approximately 4 feet 11 inches, and very slender. She spends a great deal of time in the back rooms of the shop and only performs pedicures for customers sitting in the spa chair. All five employees smile often, laugh with each other and their customers, and take time to ask about their interests, jobs, families, and so forth. I witness a very relaxed, easygoing atmosphere during the observations.

The Action

On January 21, 2009, I walk into Cutie Salon at 3:10 p.m., where I am immediately and graciously greeted by Employee 2. After a brief conversation with her, I walk over to the owner, Employee 1, who is wearing a surgeon's mask while filing a customer's nails, and request his permission to continue my observations. He removes his mask and says, "Whatever you need," in broken English. I thank him and walk back to the front of the salon to the four chairs in the waiting area. I move the far right chair and place it adjacent to the first haircut station on the left-hand side of the salon. At 3:22 p.m., a White male in his late 30s, approximately 5 feet 10 inches, and a White male child, approximately 7 years old, walk into the shop. The older man is dressed in a dirty white Florida State T-shirt; light

(Continued)

(Continued)

tan, stained, cargo shorts; and white sneakers. A black visor covers his head, with sunglasses perched on top of the visor. Landscapers use these special sunglasses to protect their eyes. His left eye is severely disfigured, with only white space showing. The gentleman's facial hair is light brown and forms a goatee around his lip and chin. A poor complexion with leathery, pockmarked, dirty skin may suggest this person works outside.

With his head down and a scowl on his face, the young boy rests his arms on the back of the chair and refuses to sit down. Employee 2 gently asks the boy to sit down and asks the man how short she should cut the boy's hair. "Shave it," he answers. "I'm serious. Maybe he'll drop his attitude then." After a few more gentle attempts, the boy hesitantly sits in the chair as the man sits in the next chair, crossing his arms. Employee 2 places a large polka-dot apron with purple, blue, silver, and black circles over the boy's torso and begins to cut his hair.

After a few minutes, Employee 2 says, "Don't move. If you move, I can't cut your hair straight." Despite her warnings, the boy continues to fidget, sniffles, rubs his nose, and scratches the back of his head. He looks at me several times with his big, round, dark eyes but never speaks or smiles. Employee 2 uses the trimmer to shave the sides but switches to scissors to cut the rest of his hair. About 5 minutes later, she asks the man if the haircut is short enough. The man never smiles and only nods in reply. After a pause of 30 seconds to switch chairs and get his own haircut, the man says, "Same thing."

During the older man's haircut, the young boy picks up one of the trimmers, turns it on, and begins rubbing it on his face. When I look at him, he has a big smile. Once the man sees what the boy is doing, he says gruffly, "No, you'll start growing hair if you start doing that."

The young boy replies, "Daddy, I'm hungry."

"We're going home soon, and I'll let you grill the steak."

Another few minutes go by, and the man's cell phone rings. He answers it using the speakerphone feature but quickly turns it off when he sees me look at him. The man says, "This better be the last time, because you're a lot smarter than that, and you're not using your judgment at all. You just tell Kathy you need her home, and that's it."

While the man's on the phone, the boy continues to pick up combs, scissors, and other objects from the counter.

"Quit getting into everything," yells the man. "No, no, don't pick them up, please. No, don't cut anything."

Employee 2 exclaims in broken English, "Those scissors are very sharp. Don't touch."

The man continues, "I know you're bored. Pick your nose or something."

Even after these requests, the boy picks up a comb and begins combing his hair. With an angry tone, the man says, "Zack, will you stop? I'm not going to ask you again."

Whining, the boy replies, "I'm cleaning off the hair," and continues to use a brush to remove the freshly cut hair from the man's shoulder.

Finally, Employee 2 finishes the man's haircut. He and the boy move to the front of the counter, exchange money, and the man says to the boy, "Let's go, dude."

At 3:51 p.m., a middle-school-aged White female walks in the salon with a White female in her mid-40s. The teenager is slender, is approximately 5 feet 5 inches, and has blue eyes, a bulbous nose, braces, and dirty blonde hair. She is wearing a pink T-shirt with half sleeves adorned with a silhouette of Barbie and the word *Barbie* written over the left shoulder. A large, silver necklace with a heart pendant hangs from the teenager's neck. Her blue jeans have embroidered designs on back pockets, and her sneakers are decorated with checkered boxes with small, black trees.

As soon as the teenager sits down in the third hairstyling station, a White male in his 30s walks into the salon, carrying three raw shrimp on white paper towel. Without hesitation, the man walks toward Employee 2 and says, "$5 a pound for 10 pounds or more. Less than 10 pounds is $6 a pound." Employee 2 looks at the shrimp and asks the man to repeat the prices. He repeats the prices to Employee 2 and proceeds to raise his voice to make a general announcement to everyone in the salon. No one in the salon expresses any interest in his offer, so he leaves. He is wearing black rubber wader boots, a camouflage jacket, a baseball hat, and dirty jeans. After his departure, the entire salon smells of raw fish.

At 3:56 p.m., a cell phone rings, and the older woman who accompanied the teenage girl answers the phone. She speaks English with a German accent but eventually switches and speaks entirely in

(Continued)

(Continued)

German during the phone conversation. As the woman speaks on the cell phone, the radio begins playing the song "99 Luftballons" by Nena, a song with German lyrics. This is an amazing coincidence.

At 3:58 p.m., a young man, approximately 16 years old, walks into the salon. He is of Asian descent with spiked black hair, wears long shorts below his knees, with a long, silver chain extending from his front right pocket, and wears a red short-sleeve shirt and white sneakers. He immediately walks behind the counter, bangs on the counter with his left hand, and proclaims, "Bye, going to the mall."

Employee 2 responds in an angry tone, speaking to him in Vietnamese. The teenager leaves the salon without making eye contact with anyone or replying to Employee 2's outburst. A customer asks Employee 2 what she said to the teenager.

"He has no job. Why are you going to the mall? Today you have job; tomorrow you have no job." As soon as she finishes translating her comments, a new song begins playing on the radio, "Bad Day" by Daniel Powter. This is another strange coincidence.

At 4:03 p.m., the teenager sits in the chair as Employee 2 pulls back her hair, parts it down the middle, and places a white clip to hold the hair in place. Employee 2 folds 2-inch-by-2-inch squares of tinfoil into the teenager's hair. A total of eight tinfoil squares are placed in various locations around her scalp. The older woman sits next to the teenager, still talking on the cell phone in German. In the meantime, another woman is speaking Spanish to two small children. A man meets the children and the woman at the front counter and begins speaking Spanish to the woman. He pays $45 to Employee 1, and all four people leave the salon.

At 4:08 p.m., a White male in his mid to late 50s enters the salon and asks Employee 3 for a haircut. His salt-and-pepper hair looks dirty as it falls over his ears and the back of his neck. He wears khaki pants, a long-sleeve black fleece, black sneakers, and black socks. His face looks weathered, with crow's-feet at the corners of his blue eyes and full, bushy eyebrows that extend over his eyelids. When he smiles, I notice he is missing two upper teeth. In a slow, Southern drawl, the gentleman says to Employee 1, "How you doing, buddy? Just a regular haircut. My boss is fussing at me. Appreciate it." Employee 1 asks the man where he works. "I work at Albertsons. Great bosses over there. I can ask my boss anything, and she'll do it for me. Worked there 7 years."

After a few minutes, he sees me looking at him and says, "It's pretty long, ain't it? I've been so busy; I ain't had a chance." I nod to him in response.

Using a trimmer, Employee 1 starts with the side of his customer's head and cuts his hair above and around his ears. The gentleman sits quietly and stares in the mirror as Employee 1 proceeds to use the trimmer to cut hair. The man presses his lips together with a slight frown. Sometimes, he closes his eyes, especially when Employee 1 gently pushes his head down to trim the back of his neck. A few minutes later, the man makes eye contact with me again, smiles, and raises his eyebrows three times in a row. "My hair was long," he quips. I nod in response.

It is 4:18 p.m., and I can still smell the pungent odor of raw fish. The teenage girl moves to the back to sit in one of the two chairs with a built-in, plastic, dome hair dryer positioned next to a watercooler. She sits down but does not pull down the hair dryer mechanism.

At 4:21 p.m., Employee 1 trims the man's bushy eyebrows. This is an incredible transformation. It looks like he is 15 years younger. Employee 1 sprays the top of the man's head with water while he combs his hair. The song, "Wanted Dead or Alive," by Bon Jovi blares on the radio as Employee 1 places hunks of hair between his middle finger and index finger and snips away locks of hair. This process repeats over and over until Employee 1 switches back to a trimmer at 4:29 p.m. This time, the trimmer cuts the sides, around the ears, and the back of the neck. Eventually, Employee 1 removes the apron and uses a hair dryer to blow excess hair away from the neck, apron, and the man's legs. After he puts a bead of gel in his palm, Employee 1 rubs his hands together and massages the man's scalp. After a few quick strokes with the comb, the man gets up to pay at the counter.

At 4:34 p.m., a White male sitting in the waiting area says to the man paying Employee 1 at the front counter, "They won't recognize you," and laughs. He repeats the sentiment again, laughs, and says to Employee 1, "It's the way I want mine right there."

Once the transaction is finished, Employee 1 invites the man sitting in the waiting area to sit in the first hairstyling station. He is in his mid-50s, stands approximately 5 feet 10 inches tall, and weighs 215 pounds. His double chin shows a five-o'clock shadow on his wrinkled, pale face. In stark contrast to the previous client, his eyebrows are

(Continued)

(Continued)

barely detectable. Crow's-feet stretch out of the corners of his olive green eyes, and freckles, age spots, and blemishes dot his face. He has mostly gray hair, yellow teeth, and very long eyelashes. This man wears a button-down, navy and light blue short-sleeve shirt with a pocket over the left chest, blue jeans, and brown loafers without socks.

A woman sitting in the spa chair, soaking her feet, shouts across the salon to Employee 1, "He wants it real short back there." The man does not react to her comments and looks straight into the mirror. About 10 minutes later, the woman sitting in the spa chair calls out again, "Shorter on the top."

This time, the man responds to Employee 1 and states, "She usually cuts my hair. But she doesn't know how to use the clippers."

Employee 1 whispers quietly, "Don't worry; I'll take care of it."

The man replies, "Whatever my wife wants," and giggles.

At 4:35 p.m., the teenage girl moves back to the seat in front of the mirror. Employee 2 removes all the tinfoil from the teenager's hair. A blow dryer and pick comb provide the finishing touches, and the teenager walks with the older woman to the front counter to pay for the service.

At 4:53 p.m., a faint smile appears on the man's face while his eyes remain closed. Meanwhile, Employee 1 uses a trimmer to trim the sides around his ears and the back of his head and neck. The customer appears to be in a hypnotic, trancelike state as Employee 1 trims around his head. The man suddenly opens his eyes and briefly watches in the mirror. While this occurs, Employee 1 looks behind him and calls out to the woman sitting in the spa chair, "Is the back short enough?"

"If my wife likes it, I like it. Looks good. It'll grow back out in two weeks," said the man. "When I was a kid, this is the kind of haircut I used to get."

"If you want it more short, I do it more short," replied Employee 1.

"No, that's good," reassured the man. "My wife usually cuts my hair; it grows back."

At 5:03 p.m., Employee 1 wets the man's hair and begins using thinning shears along with a comb. He lifts the hair with the comb and cuts rapidly on top of the scalp with the thinning shears. Employee 1 asks the man for his approval.

"Yup, just comb this down right here," he points to the front of his hair over his forehead. Employee 1 complies with the man's request, removes the apron, trims around his neck, uses a blow dryer to remove excess hair from his shoulders and lap, and finishes the job. The man thanks Employee 1 and remarks, "My wife will pay when she is done."

About 4 minutes later, the man's wife walks over to him and he says, "How you like it?"

"Feel like you had a load lifted?" she asks.

"Yeah, I feel lightheaded," he jokes.

"Well, that's what you wanted."

At 5:10 p.m., I begin packing up my gear to go home, when on the radio I hear "Home" by Daughtry. Another coincidence? Check that. I'm going to buy a lottery ticket.

All the Time in the World

If I had all the time in the world to study the Cutie Salon, I would look closely at the interactions among the five employees. A careful discourse analysis of their interactions may reveal some rather interesting characteristics of their relationships with one another. For instance, they spend a great amount of time asking their customers questions about their days, their families, and life in general. Do customers ask the employees questions about their past and present life experiences? As Vietnamese immigrants in their 30s and mid-40s, Employees 1 through 5 probably have fascinating stories to share. What are the relationships among the employees? Are they related, married, family? A longer study may reveal answers to these questions.

The clients themselves deserve intense observation. I am amazed by the diversity among Cutie Salon's customer base. Their clients come from all walks of life, spanning all age groups, ethnicities, economic statuses, and genders. A well-dressed man in a suit sits right next to a man wearing ripped shorts and a dirty T-shirt, waiting to get a haircut. Furthermore, customers and employees spoke in four different languages during my short observation. How does Cutie Salon manage to attract such an eclectic clientele? Despite this diversity, what do their customers share in common that brings them to this place of business?

Finally, with all the time in the world to study this place, I would spend time taking before and after pictures of the patrons. For example,

(Continued)

(Continued)

I wish I had before and after photos of the gentleman who works at Albertsons Supermarket. His transformation in 30 short minutes was quite remarkable. He also seemed to carry himself differently after the haircut, especially when he received positive feedback from the other customer. The gentleman seemed refreshed, uplifted, and healthier after his haircut. A photo analysis might provide valuable insight into this phenomenon. Speaking for myself, I know I always feel better after a haircut. Of course, a photo exposé may reveal the opposite occurrence, especially when a customer is angry with his or her results. Comparing employee and client discourse and body language along with before and after photos would provide me with more data to analyze in order to detect patterns and themes. In the end, the Cutie Salon is an intriguing place to observe.

Self-Evaluation

This exercise challenged me in several ways. A great deal of action and interaction occurred at the Cutie Salon. It took a great deal of practice to learn how to focus on specific events and ignore others. In some respects, I gave myself permission to tune out stimuli and filter what was coming in. For example, most of my observations focused on the left side of the salon where customers received haircuts, hair coloring, and styling. With five hair stations, I found it challenging to track the episodes at every station. Therefore, I focused on one station closest to me in proximity. This filtering process allowed me to dig deeper into my observations rather than scratching the surface on multiple data sources. Because most of my time was spent on the left side of the salon, I was unable to observe manicures and pedicures. I wonder if unconsciously I stayed within my comfort zone of studying what I know—haircuts—versus what I am unfamiliar with—manicures and pedicures. In other words, gender bias may have played a role in my decision to stay on the left side of the salon, because most men do not receive pedicures or manicures. In hindsight, I wish I would have switched sides and completed my observations on the right side of the salon.

I approached this assignment with great interest because I am a customer of the Cutie Salon. Employee 3 has cut my hair for almost 2 years, and I am impressed with his skills and the entire salon's level of service. Walking around with a camera was a bit awkward at first, but I managed to capture some interesting pictures during my initial observation. This experience set the groundwork for future observations because I became a guest in their business rather than a paying customer. The idea of *guest* was always in the forefront of my mind, because I did not want to overstay my welcome by making the employees or patrons feel uncomfortable. After all, the entire store is only 1,200 square feet, so I stuck out like a sore thumb. Remaining inconspicuous was nearly impossible.

Thus, staying conspicuous created some difficulties in the field setting. There were times when people caught me staring at them as I was typing on the computer. I was pleasantly surprised none of the customers approached me to inquire about my venture. However, there were a few times when I came close. I was definitely in the radar of the gentleman with the disfigured eye, whose son was playing with the different beauty products. There were a few uncomfortable moments as he watched me while I observed his son and typed notes on the computer. I tried to avoid eye contact from that point forward. Another difficulty I encountered was learning how to stem the urge to participate when I was supposed to be a silent observer. I constantly found myself fighting the impulse to engage in conversations that occurred all around me. This was when focusing on one single event became critical, as it helped me focus rather than become distracted by outside stimuli. Remaining focused prevented me from participating in the events as they unfolded before me.

As a researcher, I learned that I am much more observant than I gave myself credit for in the past. When I take the time to stop, watch, and listen to what is happening, I gather an incredible amount of data. Painting a picture with words is likely, especially when I take the time to reflect on what I am observing. I learned that this reflection piece is absolutely critical for me because it forces me to stop and think about my observations. My professor suggests journal writing as a means to reflect on the data as well as "refine ideas, beliefs, and my own responses to the research in progress." Rushing this process is a mistake, since important pieces of information are lost. I took my time, focused on specific events, and learned that, as a researcher, I can collect valuable and relevant information in the field. This assignment was good practice for future field work.

Exercise 3.2

Reflecting to Strengthen the Writing Habit

Important Life Event

Think of an important event in your life or in a project you are writing. Write the weather for the day it happened. Try to recall what you were wearing. What was happening around you? Write a description of that event and all the subsequent meaning in the event. Aim for five pages to start. Share your writing with another person and get feedback. Rewrite. Add this to your researcher reflective journal as part of understanding your role as a researcher and reflective agent.

Exercise 3.3

Writing Your Educational Autobiography

Write 7 to 10 pages of your educational autobiography with particular attention to your career choice as an educator and how you are progressing on your educational journey. Think of this as your learning autobiography. What inspired you to be a teacher or leader? What influences or setbacks have you experienced? If you could do this all over again, would you still select education as your field? Add this to your researcher reflective journal as part of understanding your role as a researcher. For those who are in other professions, it is always good to reflect back on your educational autobiography. Add a few questions, such as the following: Are you still in the same field in which you were prepared? What outstanding influences did you have during your educational matriculation?

Exercise 3.4   ■

Writing a Pedagogical Letter   ■

The Pedagogical Letter: Write a 7- to 10-Page Letter to Someone You Know

One of Paulo Freire's (1921–1997) favorite formats of communication was through letters. He described in these letters his political, sociological, ideological, philosophical, and contextual beliefs, values, and ideas. Most of his final text, *Pedagogy of Indignation,* evolved and is generated from these pedagogical letters over the course of his lifetime. These letters contained his hopes, his emotions, and his sensibilities. This exercise is for you to do the same.

Write a letter and explain your critical pedagogical beliefs regarding an educational issue you care about and how you apply these beliefs in your everyday world. For example, some students have written to their own children, their grandparent, a spouse, a significant other, another student, or some historical figure, like Mother Jones or Eleanor Roosevelt, or someone who has inspired the writer. Some have written to Paulo Freire himself or one of the authors of a favorite text. Write in a letter format, and include this in your researcher reflective journal.

■ Next Steps

The next three protocols you will see show some possibilities for becoming an active agent as a researcher. The learner takes responsibility for self-evaluation, as in the sample Figure 3.1, Self-Evaluation for Your Journal. A second option, Figure 3.2, My Story to the Best of My Knowledge, allows for a more introspective turn. Finally, Figure 3.3, Model Format to Explain Your Qualitative Research Study, is to be seen as a working document and a starting point for the learner to begin constructing the ministudy that brings together the practice of and learning from all of these exercises to this point. Now, the learner moves on to the cycle of interviewing. The following self-evaluation is an option for those students who prefer letter writing over the sample forms I have constructed.

Figure 3.1 Self-Evaluation for Your Journal

Self-evaluation of my _____ activity.

List three adjectives that describe what you learned from this activity.

1.

2.

3.

Locate Yourself

Historically, the theories that have most affected me and shaped me are: (List at least three authors). EXPLAIN.

1.

2.

3.

The reason these authors have shaped my thinking are:

Figure 3.2 My Story to the Best of My Knowledge

In this exercise, I want you to reflect on your intellectual growth and development. Which ideas have dazzled you? Prompted you to go further in your studies?

Try to remember at least one to three incidents from school, kindergarten to the present, that have profoundly affected your thinking. Describe in detail these incidents—sights, sounds, smells, tastes, the feel of them, key participants, your own memory of how you reacted at the time, and how you view these incidents today. Add this entry to your portfolio.

Name _____ Date _____

Incident One:

Incident Two:

Incident Three:

You may not see what is in front of you. You may look at something, but you may not look carefully enough to see what is there.

Figure 3.3   Model Format to Explain Your Qualitative Research Study

1. The purpose of this study is to describe and explain: (example: the mentor's view of mentoring)

2. The theory that guides the study is: (example: phenomenology)

3. The exploratory questions which guide the study include:
   a. What elements or characteristics make up this mentor's beliefs about mentoring

   b. What variables influence this set of beliefs?

4. The literature related to this study includes:
   a. Methodology literature (name the area)

   b. Topical literature (name the area)

My rating involves:

I still want to work on:

■ **Sample Reflection (Edited) for Adding to the Reflective Journal**

Self-Evaluation Upon Observing in a Student Laboratory for English-Language Learning

By Oksana Vrobel (2009)

Doctoral student in second language acquisition and instructional technology

Challenges on the Way

I am glad that I have done the observation assignment. It was useful and challenging simultaneously. Because I am a beginning qualitative researcher, these are my first steps and attempts to collect data, analyze them, and self-reflect. As many suggest, practice in data collection, analysis, and writing is essential. Therefore, I appreciate that I had this opportunity to practice and receive a valuable feedback on it. Despite the unquestionable benefit of this exercise, it was challenging. This observation work required absolute commitment and concentration. I am a perfectionist by nature and try to do my best. Therefore, it took some time to think over the task itself, to plan how I was going to approach it, to do the observations, and, of course, to write the observation report.

One of the challenges that I had was the time I chose to do observations. During the first time when I came to observe a setting, the place was empty, and I had known that it would be empty because it was just the beginning of the semester in the English Language Institute (ELI). The lab was still closed. I had enough time to look at the details of the setting. However, when I went there for the second and third times, I had to evaluate the situation to determine whether it was the best time for observations. There were only three students there, and they were leaving. So, I decided that I should choose a different time to complete the second and third observations. I know that planning observations is also one of the skills I need to practice, but at the same time, there are conditions that I cannot predict. Thus, one of the lessons that I learned was always to leave some time for a back-up plan in case I could not do an observation due to some unexpected obstacle.

In addition, it was rather hard to focus on the details of the setting and people's appearances. I think that it is just one of my traits of character; I am always absorbed in my inner world and not used to focusing on small details that exist around me. Though I realize that small details, such as a hand position while writing, may tell a researcher some essential information about a participant of a study, it was hard to focus and go over the entire setting. While doing it, I just remembered that it is always hard the first time. I also did not know what details I needed to pay attention to most. So, I just followed the rule, "The more I can describe, the better it will be." I think it is possible to think about some details that can be crucial to observe. However, I need to observe and describe many nitty-gritty things that, as a result, can provide important information.

Moreover, I found some internal resistance that made me hesitate, whether it was the best time for observation, for self-reflection, and for writing the report. I think this resistance is the result of my inexperience. I need to overcome the resistance with constant practice. I agree with those who think that it is always possible to find something that seems more important than writing. Even such a humble thing as a dishwashing can seem to have importance when we try to block the resistance to write and practice constantly. I found I need to work at writing.

Approach to the Assignment

I think I clearly see this assignment as a means to practice and have a hands-on experience in observation. I have never done an observation before. Therefore, I carefully planned and tried to think over what I needed to do for an observation, for example, times and stages of it. I tried to plan it at home. However, when I came to the ELI, I still had to take a table to jot my observation notes down. I did not use a laptop to put down notes. I thought that, for the first time, it would be more convenient for me to use a pencil than to type and organize the information at home. I think that next time I will bring my laptop to put down notes. I wonder what could be the most effective way to write down notes and what literature says about it. My guess is that it depends on each researcher's preference. Later, I discovered my own style and preferences.

After each observation, I tried to write and organize my notes to be more prepared for the next observation. I think doing it step-by-step

(Continued)

(Continued)

was a good decision. It makes the process more planned, organized, and meaningful for me. It also gives me more time to focus on each of the stages separately instead of having just a holistic overview of the experience. Fortunately, I had an opportunity to take pictures of both the setting and one of the students I observed. I asked for permission of both the administration and the student, and they allowed me to take pictures without hesitation. I think that I was lucky because the person whom I observed during my second observation was a student of mine two semesters ago. In my opinion, this contributed to her eagerness and quick agreement to be photographed. However, I think it could be challenging if the person were a stranger. Then, it could raise suspicions and unwillingness. I definitely have to keep this in mind if I am going to observe strangers for my future studies. This is also applicable to the setting. I chose the place because I feel that I have some personal connection to it as an international student and an instructor. However, it is also a convenient sample simultaneously. The ELI is my place of work now, and I had no problem getting permission from them to observe it and take pictures of it.

As for the writing, it is my first time, and I am not writing a paper part by part. This time, I tried to focus on my inspiration and internal readiness to write different aspects of this assignment. I think this nonlinear approach to writing is a little bit confusing at first, but I was so inspired by the book *Writing Down the Bones* by Goldberg (2005) that I decided to follow her advice to start writing about anything that I was up to at the moment. In my opinion, this experiment was a success; I enjoyed it more, though I think the process of editing will be more time consuming this time. While writing, I had to stop to think then to write, teach, read, and write again. It was not an easy process. Writing always requires a lot of effort. I am exhausted after it, but it is self-rewarding at the same time.

Difficulties in the Field Setting

In addition to the challenges described in the first part of my self-evaluation, I think it was rather difficult to do the third observation. I needed to observe people in action and to take notes about everything around me. However, it was extremely hard because there were several groups of students in the lab, and when I focused on a group, I could not take notes about the others. Moreover, I felt that even when I took notes and got distracted for seconds to write, something important might happen, and I would not see it. I realize that this was just a practice, but I wonder how experienced researchers deal with this question and whether they have this concern at all.

As I have mentioned above, I did not have any difficulties with getting permission to observe a setting and to take photos of it. However, I still felt uncomfortable observing people and their actions. I did not want to make them think I was staring at them. This could have caused misunderstandings and changes in their behavior. In the worst case, they could have left. Therefore, it is important to think about possible difficulties that can occur at the setting ahead of time and eliminate them if it is possible.

Second Chance

If I were to return, I think I would try to use a laptop to take notes and see if it is more effective than to take notes in pencil. I would also try to record my oral description of the setting and people. I think I need to try different approaches to observations to be familiar with the ways to do it. In addition, I will learn what I prefer. I understand that the purpose of this assignment is to practice and take the first steps in qualitative data collection and analysis. Having this experience now, I can try to do observations as a part of my future qualitative studies.

Lessons Learned

I think it was a valuable experience for me. Though I read a lot about data collection for class, there is nothing better than to experience and practice data collection myself. During observation, I learned that I need to focus on the details. It was challenging, but I think it will become easier with practice. In addition, I experimented with different approaches to writing and self-reflection. I tried to write in a nonlinear fashion. That was worth doing.

In addition, I concentrated on my own experience as a researcher. I think this is one of the advantages of qualitative research. A

(Continued)

(Continued)

researcher's experience and reflection are essential in qualitative studies. A qualitative researcher is an instrument. A qualitative researcher designs a study, collects data, analyzes them, and provides a rich description through his or her own experience and background. I think it makes a lot of sense, because any study that I design is a part of me. I can do a study and clearly describe connections between this study and me and my role in the study, not only in the rationale but also throughout the paper.

Furthermore, I discovered how much I enjoy the whole process. Sometimes, when I am doing an assignment, I do it for the purpose of simply meeting the requirements of the course. This assignment is not among those. I think I enjoyed every minute of doing it and writing about it. Thus, observation of the ELI computer lab, students, and their actions led to the conclusion that ELI students use the setting to study individually and in groups, socialize with their friends from other countries face-to-face and online, keep in touch with their families, interact, and use the resources available. Part of their life experiences, emotions, and feelings is invisibly imprinted in that setting.

The observation assignment was beneficial for me as a future researcher because it provided me with a precious experience and practice. I hope that now I can plan to include observations into my data collection for future qualitative studies.

■ **Pitfalls**

Learners who have the luxury of a class situation or a community of scholars with whom to discuss daily progress have remarked on the following pitfalls to these observation exercises:

1. *Focus and concentration* are the critical elements to doing a decent observation. The first obstacle to overcome is deciding what to observe, because you cannot possibly observe everything in a given social setting. The best way to approach this is to settle on a section of a setting, one person, or one set of objects to begin observing and describing. Many describe a feeling of frustration as they have never observed closely before. That dissipates after some practice.

2. *Overdoing it, that is, overdescribing one piece of a setting, for example, may hamper the end result.* Try not to fixate on just one piece of the scene. See if you can gradually add more to your observation as you go along. Once, a student wrote 20 pages while describing a flute. This is an amazing feat and could have even been enriched by placing this flute in a social context with description of the flute player, the actual setting where the flute was being used, and how this observer came there in the first place.

3. *Underdescribing* could harm the overall effect and goal of the exercise if not enough attention is paid to each component. When in doubt, keep writing and refining those observation skills. Take baby steps at first and then move on.

4. *Interacting and communicating with on-site personnel* in the field can be a problem. When entering a public space and in trying to get permission to take photos, realize that some places will allow it and some will not. It is a good idea to explain you are taking a course, and this may help you a great deal. About 90% of people who ask for permission to photograph a setting, for example, get that permission and often a free coffee. As a backup, should you not get that go-ahead signal, try drawing a floor plan and sketch the scene as best as is possible. Use narrative description as a rule; the better you become as a describer, the less you need to photograph. Also, a note about photographs here is helpful. Photographs should be taken if they add something to the study. They must be purposefully selected for your narrative report. They should add cohesion to the report. It does not make much sense to throw in a dozen photos without a narrative, cohesive reason for doing so.

5. *Use diligence* as you approach these exercises, and take the time to be reflective. As you may already realize, this type of research cannot be rushed. In our overscheduled, overtexted, over–e-mailed existence, we often lose site of the value of silence, reflection, and tapping into our own creativity.

■ Summary of Chapter 3:
Advancing the Observation and Reflection Habit

In this series of four exercises, the learner has direct experience in advancing further in observational techniques by visiting a public setting three times and describing, explaining, and interpreting that setting for a reader. Next, the habits of reflection and writing are reinforced and advanced through the practice exercises of writing a pedagogical letter, an educational autobiography, and describing and reflecting upon a critical life event. The exercises move forward with increasing complexity in sharpening one's habits of observation, reflection, and journal writing. Each learner is encouraged to evaluate and document in a research reflective journal as he or she moves through the exercises. For each individual activity, the learner then completes a summative evaluation statement on his or her own progress overall as an observer. These exercises are in preparation for progressing to the next series of exercises: interviewing and writing. Furthermore, these exercises have their origin in the arts and humanities. Many are used for the purpose of developing the narrative writing habit.

Because of the hectic pace of many of our lives, being still long enough to observe and describe a setting, an object, or a person requires us to settle down and let ourselves become totally absorbed in the given activity. Most learners seem to get better at this the more they do it. In this case, practice really does allow the individual to improve and proceed in the series. Like the dancer in training, who improves through consistent, relentless, and disciplined exercise, the qualitative researcher in training may improve as an observer of the human condition. Like the student of yoga, the more you concentrate on a given posture and breathe into it, the better you will be able to train your body to hold it. Like the professional writer, the writing habit needs to be developed, and that is best done by writing every day. As we move to the next section on developing the habit of interviewing and writing, digital exercises will also be added to practice and refine skills associated with working on the web and for the purpose of critical development.

■ Resources for Understanding Observation

Denzin, N. K., & Lincoln, Y. S. (2003). *The strategies of qualitative inquiry* (2nd ed.). Thousand Oaks, CA: Sage.

Edwards, B. (1999). *The new drawing on the right side of the brain.* New York: Putnam.

Eisner, E. (1991). *The enlightened eye.* New York: Macmillan.

Freire, P. (2004). *Pedagogy of indignation.* Boulder, CO: Paradigm.

Goldberg, N. (2005). *Writing down the bones.* Boston: Shambala Press.

Leavy, P. (2009). *Method meets art: Arts-based research practice.* New York: Guilford.

Merriam, S. B. (2009). *Qualitative research: A guide to design and implementation* (Rev. ed.). San Francisco: Jossey-Bass.

Saldana, J. (2009). *The coding manual for qualitative researchers.* Thousand Oaks, CA: Sage.

Stake, R. E. (2010). *Qualitative research: Studying how things work.* New York: Guilford.

Strauss, A., & Corbin, J. (1990). *Qualitative research methods.* Thousand Oaks, CA: Sage.

Sze, M. M. (Ed.). (1997). *The mustard seed garden manual of painting.* Princeton, NJ: Princeton University Press.

Van Manen, M. (1990). *Researching lived experience.* New York: SUNY Press.

Wolcott, H. F. (2001). *Writing up qualitative research* (2nd ed.). Thousand Oaks, CA: Sage.

CHAPTER 4

The Interview and Writing Habit ■

History is nothing but a series of stories, whether it is world history or family history.

—Bill Mooney and David Holt (1996)
The Storyteller's Guide

Interviewing has taken on a new tone recently with Internet inquiry and interviewing individuals virtually, that is, on the Internet and on websites. As a result, many are wondering what will happen to the tried-and-true, face-to-face interview. Usually, the most rewarding component of any qualitative research project is interviewing, and it will never be replaced or never fade away. Yet, the question of whether we still get good data virtually remains. What are the differences in face-to-face interviews and interviews on the World Wide Web? In this chapter, I will discuss these points, but I begin with learning about traditional, face-to-face interviews. In addition, I will discuss the issue of how to present interview data, and you will find narrative and poetic examples in the form of found data poems, that is, poetry found in the transcripts of the interviews. Interviewing increases your skill set and mind-set as a qualitative researcher, building on your observation skills. Whereas observation is the act of taking notice of something, interviewing is an act of communication. In fact, a major contribution to our history as qualitative researchers is the growing literature on solid interviewing techniques. Since the 1970s, more articles and books have become available (see Janesick, 1991; Kvale, 1996; Kvale & Brinkmann, 2009; Rubin & Rubin, 2005; and Spradley, 1980). I consider this a tremendous leap forward, because interviews provide such rich and substantive data for the researcher and are also a major part of qualitative research work. Although I have written earlier about the importance and ways of approaching interviews, I would like to frame this chapter in terms of some of those earlier key points, which I will summarize here. A good deal of what I have learned about interviewing ultimately has come from trial and error within long-term interview studies. This chapter is meant to be a nonthreatening and systematic way to approach the complex and challenging act of interviewing another person in person, face-to-face. In addition, we will examine the virtual interview in its complexity.

■ Two People Talking, Communicating, and Constructing Meaning

Interviewing is an ancient technique, and for the purposes of this text, I define it in this way:

> *Interviewing is a meeting of two persons to exchange information and ideas through questions and responses, resulting in communication and joint construction of meaning about a particular topic.*

As we are always researchers in the process of conducting a study, we rely on different kinds of questions for eliciting various responses. In fact, one of the solid reasons for doing a pilot test of your interviews through a pilot study is so that you, the interviewer, learn which questions are best suited to your study, under what conditions, and when to use particular types of questions. In addition, you may learn how to probe further in a semistructured interview situation as participants tell you their stories. Writers like Spradley (1980), Mishler (1986), Stake (1995), Rubin & Rubin (2005), Berg (2007), Kvale & Brinkman (2009), Merriam (2009), Roulston (2010), and others suggested types of questions for interviewing that have always worked for me and my students. I categorize these as basic, descriptive, big-picture questions; follow-up questions; comparison or contract questions; specific example questions; and structural questions, and overriding all questions is the clarification question. I expand these notions and offer the following as examples of types of interview questions.

■ Types of Interview Questions

Basic, Descriptive, Big-Picture Questions

Can you talk to me about your life after Hurricane Katrina? Tell me what happened on that evening.

Describe how you felt that evening.

Of all the things you have told me about being a critical care nurse, what is the underlying premise of your workday? In other words, what keeps you going every day?

Follow-Up Questions and Clarification Questions Including Example and Experience

You mentioned that you have no one to talk to about your reactions to the hurricane. Can you tell me some of those reactions?

You mentioned that you loved going to New York. Can you give me an example or two of what made you love your trips to New York?

Talk about your impressions of the city.

You have used the term *socioscientific issues* today. Can you clarify that for me? What exactly can you talk about regarding your teaching of science through socioscientific issues?

Comparison, Contrast, or Structural Questions

You said there was a big difference between a great professor and an ordinary professor. What are some of these differences? Can you describe a few for me?

■ Preparing Questions

A good rule of thumb for interviewing is to be prepared. Compose as many thoughtful questions as possible. It is far better to be overprepared than to get caught in an interview without questions. Usually five or six questions of the types just described are reasonable and may yield well over an hour of interview data on tape. A simple question like, "tell me about your day as a cocktail waitress," once yielded nearly 2 hours of interview data, leaving all the other questions for another interview time. You will learn to develop a sense of awareness about your participant(s) in the study and rearrange accordingly. Also, always get permission for your interviews in writing. Increasingly, Institutional Review Boards (IRBs) are demanding more of qualitative researchers in their consent forms. See Appendix F for a sample form that has been accepted lately at my university. In order to prepare for testing out some of the questions you create, let us proceed to some exercises that will give you some experience with interviewing. In this section of the text, all of the interview exercises will use the following format:

1. First, be prepared with a digital tape recorder, back-up recorder, and a notebook to take field notes while interviewing. You may also wish to use the Livescribe pen, which works with particular notebook computers either in written or speech modes. Some interviewers prefer a digital video recorder to upload the interview to their TV sets, computers, YouTube, or their own websites. This gives plenty of time for physically seeing the interview take place as one transcribes the interview. Note that if you send your interview to be professionally transcribed, most companies prefer digital voice and video recorders rather than the previous generation of tape recorders. One hour of taped interview yields about 20 pages of interview transcript. On the open market, the approximate cost for a transcriptionist is about $100 to $120 per hour of taped interview, so budget accordingly for your dissertations or other research projects. You will get to know your style as you move into interviewing techniques. (See Appendix I for information on digital equipment and transcription services.)

2. Before the interview, check your digital voice recorder or video recorder to see that it is functional. Test your voice on the tape by noting the *date, time, place,* and *the name of the participant* on the tape. This is helpful later, not only when you do the transcriptions of the tape but also when you need to jog your memory at a subsequent date. You may also find this will serve you later as a coding mechanism for your hours of transcripts.

3. Whenever possible, carry a back-up recorder and batteries. Many cases have been described where the recorder was malfunctioning, the recorder died, or the batteries wore out! Many newer models can be recharged and have a thumb drive attached to insert into your computer and upload your interview onto a CD.

4. If you feel more comfortable giving a copy of the interview questions to your participant, do so ahead of time. Call ahead to remind and verify the

exact date, time, and place of the interview, and arrive early. Remember, in the social world, anything can happen, and be prepared to reschedule an interview if requested.

5. Always have a back-up plan. If someone decides to leave the study, be prepared to replace that person.

6. Field test your interview questions with another member in class or someone who will give feedback to prepare you for the first days in the field. Always do a pilot test of your interview questions for your dissertation study.

■ Examples of Interview Questions From a Recent Study

Interview Protocols

By Anete Vasquez (2009), Instructor, Secondary Education, University of South Florida

From her study, *Escaping Corporate Culture for the Classroom:*

The purpose of this study was to add to the body of knowledge about what motivates individuals to give up highly successful and lucrative careers in the corporate world to enter the field of teaching without any outside forces—such as active recruitment, fast-track programs, and subsidized coursework—influencing them.

The Interviews

Three 1-hour interviews were conducted. The first interview was extremely informal and unrecorded. It occurred on Sunday, May 31, 2009, from 1:00 to 2:00 p.m. in the living room of my home. The purpose was to get some background information on Bryan. What I had intended to be a 10-minute meeting turned into an hour, because in the course of sharing some of his background, Bryan told entertaining and detailed stories of his youth.

The second interview occurred on Monday, June 29, 2009, from 9:04 to 10:10 p.m. Because I know Bryan quite well, I invited him to my home, and the interview took place at my dining room table. The interview was recorded using my iPod with a Belkin TuneTalk Stereo attachment, which was later connected to my computer to transfer the interview via iTunes and downloaded into Audacity for transcription purposes. The transcription of the interview occurred the following day, as I wanted to transcribe while the meeting was fresh in my mind. This interview was structured in an opening-the-locks pattern, where a small number of main questions are asked of a knowledgeable partner. The intent is that the questions will "be like opening the locks on a river, allowing the waters (information) to rush forth" (Rubin & Rubin, 2005, p. 144). The questions for the interview were also modeled on those in the Qualitative Research Methods in Education II course syllabus (Janesick, 2009). Questions for Protocol A included the following:

1. What were some life events that influenced your decision to become a teacher?

2. What contributed to your decision to obtain a master of arts in teaching rather than go the route of alternative certification?

3. Tell me what you like about being a graduate student.

4. Tell me what you dislike about being a graduate student.

    a. It doesn't pay as well. No, I'm pretty comfortable in an academic culture.

5. What does being a teacher mean to you?

    a. Umm, well . . .

6. Where do you see yourself in 5 to 10 years?

When I sat to analyze and code the transcription from the first interview, themes appeared immediately. One of the most prevalent was that memories of school and specific teachers had played a strong part in influencing Bryan's decision to become a teacher. Another theme was actually named outright by Bryan during the interview when he said, "Ahhh! There's something that ought to be a theme: meaningful work. People should have meaningful work." Hence, the quest for meaningful work became a theme. Along with this idea emerged the theme of the attempt to create meaningful work in the corporate world through teaching and mentoring. Another theme was the feeling of satisfaction and success at past experiences teaching or mentoring others. The interview also suggested that Bryan wanted to work with kids and that that was a motivating factor. Not only did he like the idea of teaching, as alluded to in the previous theme, but he wanted to help kids fit in and feel important and encouraged to follow their passions and creativity. Another theme was that he was motivated by a strong love of learning and the pursuit of knowledge. Dissatisfaction with the corporate world was another strong theme in Interview 1. These are the themes that emerged from the first interview:

1. Positive and negative memories of school and specific teachers.

2. The belief that everyone should have meaningful work.

3. The attempt to create meaningful work in the corporate world through teaching and mentoring.

4. The feeling of satisfaction that accompanies past teaching experiences.

5. A desire to work with kids to help them fit in and to encourage them to pursue their passions.

6. A strong love of learning.

7. Dissatisfaction with the corporate world.

The third interview also took place at my dining room table, and it occurred on Tuesday, July 7, 2009, and lasted from 11:00 to 11: 52 a.m. It was recorded and transcribed in the same manner as the second interview, and the transcription was completed that same day. I had not done an extensive review of the literature before conducting the first interview because I was afraid that, as a novice interviewer, a strongly grounded knowledge of previous research might tempt me to guide the conversation in certain ways. I preferred to conduct my first interview with a bit of naiveté on the subject, thereby fully allowing Bryan to share whatever came to his mind. Prior to the second interview, however, I conducted a relatively exhaustive review of the literature on what motivates people to change their careers to teaching. I was surprised to find that the information that Bryan had shared with me in the second interview aligned well with reasons listed in the literature. I took some of the motivations that were cited in the literature that Bryan had not mentioned and asked him specifically about whether or not these items had influenced his decision at all. The items were about financial reasons for entering the teaching profession, the influence of age or life stage on the decision to become a teacher, and major life events that may have influenced his decision. I also asked him some more pointed questions about what his conception was of the life and work of a teacher to see if his answers were congruent with the answers he gave in the second interview.

Additional themes that emerged during Interview 2:

1. Financial circumstances

2. Age and stage of life issues

3. Recent life events

Themes That Emerged in Interviews 2 and 3

Bryan is right. It does seem that, throughout his life, there were many factors influencing his desire to become a teacher, a desire that he is only now able to pursue. Numerous themes struck me as I analyzed the transcripts from our last two interviews. One of the themes is one that is commonly cited in the literature dating as far back as Lortie's (2002) seminal sociological study of teachers. This theme is the theme of past experiences of schooling influencing the decision to become a teacher.

Past Experiences of Schooling Influencing the Decision to Become a Teacher

Bryan speaks of his elementary and middle school teachers with fondness and of both positive and negative experiences that may have influenced his decision to become a teacher. Of his elementary school teacher, Ms. Altheide, he says:

> I had very good teachers at the elementary school level, specifically in first grade. Sandy Altheide was the teacher. . . . She was very loving and supportive. She made me feel welcome in the class. She made me feel like I was an important part of the class.

He also remembers a middle school teacher who had a profound impact on him for numerous reasons: He was Bryan's first male teacher, he was an athletic man, and he had high expectations for his students:

> A teacher that I really respected was Ray Viscovy, and he was sixth, seventh, and eighth grade. He was the health teacher, and he also taught science, and he was the coach—the track coach the basket ball coach and Viscovy was a real, you know, tough nuts. He had *pencilitis;* he couldn't read pencil, so you had to write in pen. He had extraordinary expectations. I mean, I remember sixth-grade health class—we had to memorize all 206 bones in the human body, and we had to put together these extraordinary

books and folders about geography and stuff, and I remember working my butt off, you know, drawing pictures of volcanoes, and magma flows, and earth's crusts structures, and all of that. It was a chance for me to draw, which was something I always loved to do, but it was in the context of, you know, putting information together, and Viscovy, by God, expected you to do your best work, and he wasn't afraid to give you an F if you didn't do your work; conversely, if you did your work, and you did your best, and you worked hard, he was your friend. I mean he was a guy who. . . . I mean, he respected you for doing that. Viscovy was a man, and he was the first male teacher who I'd had—ever—and, you know, he was a manly man, and he was the coach, and he played semipro ball, and he was the guy you looked up to, and he was a teacher, and he demanded things of you that made you work hard, and it was intellectually satisfying. . . . Viscovy was a good teacher with high expectations who was also a jock, and that was something that I wanted. I felt like a jock, and being a smart guy didn't exclude me from being a jock. So, Viscovy was a teacher who was clearly a formative influence.

Bryan also remembers an early experience in school that was not so positive but that served as a motivator for wanting to become a teacher. In this instance, a teacher made Bryan feel like an outsider because he was gifted:

In third grade, when Mrs. Irwin had spelling tests, she would segregate me from the rest of the class. She sent me to the back of the room because she didn't want people copying off of me because I always got all of them right. And so, I was the smart guy who had to be removed and put in the back because I presented a danger to others—the opportunity for them to cheat—and I didn't help people by going, "Look at my paper," but I wasn't a kid who sat over my paper and said, "Don't look at my work. Don't look at my work." I was like, "I'm doing my work, and if you look, that's your problem, not mine. I'm not gonna hoard this." Then, there were other occasions throughout my career where teachers made comments like, "He's so smart; do you think we'll ever be able to understand what he is saying?" or "Do you get the idea that he's so far ahead of us that we'll never catch up?" So there were occasions where being the smart kid was used against me. My abilities were used to distinguish me from the rest of the class and not in a positive manner.

Thus, you see here how interview questions try to get at an idea, which is what interviewing is all about. Next, let us turn to a sample of an excerpt from an interview study on technology by Carolyn Stevenson.

■ Example of a Transcript

Technology Director A: Kathy (pseudonym)

Excerpt From an Interview on Technology and Leadership by Carolyn Stevenson (2002)

Q: Let's begin by talking about leadership. How would you describe your leadership style? *[Descriptive, big-picture question]*

Technology
Director A: I would say I attempt to lead by example. I know that sounds kind of cliché, but I think that's probably true in terms of . . . not a big, hands-on manager. I should say I don't want to be big, hands-on manager. I find myself in that role probably because I'm more of a perfectionist than I should be. And I'm a little bit more of a control freak than I wish I were. So, I like things done the way I like them done, and I tend to hover over people, where I normally wouldn't like that, and I certainly don't like people doing that to me. So it's kind of hard. At the same time, what I would like is to be able to be the kind of leader that people watch what I do, do it behind me, and follow my example. I don't really have staff like that; I have some individual staff members like that but not so much of a staff like that. This is my first management job. I never managed before. So, this is pretty hard for me. It's been the hardest part of the job. I'm not a very good manager. It's because I'm a control freak. I take too many things on; that's for sure. I'm a perfectionist; that's for sure. And, I am surrounded by a culture that doesn't support a lot of change, and I'm a real change person. And, it's hard. If I were working in a change environment, then I would have people working for me who were accustomed to change. I found myself in the exact opposite. I think I did that by design. That's my whole want, to fix it mentally. So, I don't think it was by accident. But, it's hard to work in an environment where nine tenths of the people you work with are downright resistant to change in a culture that hasn't supported change, ever, in its existence when you are a change person. So, you know, I like that. I wouldn't have chosen that if I didn't, but it makes it hard. It's part of the difficulty. I find myself frustrated

by it, in terms of frustrated by people's performance, or lack thereof, in my opinion, because I think it could be done so much better, so much easier, so much faster. If it were done better, it would fix a lot of problems. I am a kind of look-at-the-root-of-the-problem kind of person instead of putting out fires, which was the way this department was running for years and years and years. Look at what's causing the fire. Solve the fire, and then sit back and be happy when you don't have to put them out anymore. [Laughter.] And that's not the way this place thinks. They don't change the root. They fix the symptom.

[Knock on the door]

      I'm sorry; I have to get that.

[Interview stops, then resumes]

Technology
Director A: So, I was saying, working in an environment where technology is the driver takes some of the personality issues out of it. Some of them, in terms of blame, blame it on the technology. Don't blame it on me because the technology can drive it. And, in the end, the technology helps it because; it supports it . . . the idea of change. When everybody adapts to the technology, and they discover that, oh my gosh, you know, it really is easier to send e-mail to all the faculty in the space of 10 minutes then it used to be to request labels from district office and draft a memo that has to be approved by everybody, and make copies of it in the reprographics department, and put them in envelopes, and put them in the mailroom. And then, it takes 10 days to get a letter out and five people involved [laughter] as opposed to sending an e-mail out to everybody in one solid swoop in the space of 15 minutes.

     So, the technology supports the change when the reality of what the technology can do becomes evident, and lo and behold, people are like, ooh, I got it. I didn't like that. I didn't want you to make me do it that way. I don't like you for making me do things differently [laughter]. But, when it comes right down to it, I can see the wisdom of doing it a different way. So, I think that's what I mean when I say leadership by example, for me anyway, and how technology has kind of, how I have used technology to my advantage to overcome some of my personal deficiencies in the way I manage people, because I don't like management. It's hard.

Q: You talked about the culture being resistant to change. Is it just because it's the philosophy that, well, this is the way it has always been done . . . or? *[Clarifying question]*

Technology
Director A: I think it's . . . again, it's a little of the chicken and the egg for me. I don't know if it's because I find myself drawn to hesitant environments that are very change resistant because I have a personality that likes to come in and turn things upside down or if every place is kind of change hesitant. And I tend to think the latter.

I think every place is . . . you know, some places more easily adapt to change than others, but when it comes right down to it, I think people are creatures of habit. And businesses in particular, larger businesses, more than 40 people, are stuck. They do things in a certain way. They have always done them in that way. They don't think about how to do things differently. They don't want to because other people are telling them they should.

So, I tend to think that most places are not big change environments. But again, I might be wrong. I might just be finding myself in big, bureaucratic organizations that are more hesitant than others. I have never worked in a new company or entrepreneurial environment. And, if I did, I might find myself surrounded by people who are more like me, and I might have a different perspective. As it happened, I have worked in government and nonprofits that are like close to glacial speed in their change [laughter]. It's coming 10 years in 2½. When I got here, this place operated as if it were 1985. And now, it operates like it's 1998. It's not really 2002 [laughter], but that's OK. It's working its way up. At this accelerated pace, in the next couple of weeks, we might actually be in the 21st century. *[Interview ends]*

Thus, you see examples of questions for an interview project and an excerpt of a completed interview, and you also have a sense of some preparation behaviors for the interview. Let us turn to some exercises to get you into the interview habit.

Exercise 4.1   ■

Interviewing Someone You Know   ■

Find someone you know to interview on any one of the following topics for 20 minutes.

 1. What are your beliefs about friendship?

OR

 2. Describe for me someone you admire, a historical figure, or someone alive today. Explain why you selected this person and why you admire this person.

OR

 3. Describe your typical workday from the moment you arise in the morning to the end of your day. Remember to end each and every interview with the question: Is there anything else you wish to tell me at this time?

Be sure to tape your interview and take field notes to train yourself to observe nonverbal cues and behaviors.

Discussion:

 1. How did you approach this exercise?
 2. What was most difficult for you?
 3. Would you change anything the next time you interview someone?

Learners now get a chance to practice working with taped interview data by transcribing at least a portion of the interview, if not in its entirety. Later, the group will practice analyzing data from these interviews in groups, with partners, or alone, as they see fit. In the context of the classroom, members pair up, and each member practices as an interviewer and an interviewee to see what it feels like to inhabit each of those roles.

(Continued)

(Continued)

Evaluation:　Learners follow the same evaluation exercises as in the previous chapter, keeping track of their progress and reflecting on the meaning of the exercise by writing in the researcher reflective journal.

Exercise 4.2 ■

Interviewing a Stranger ■

Learners follow the same directions as above; only now, they must find a stranger to interview. In a class setting, learners go out on campus and find someone. If trying this on your own, use your imagination to find someone in either the workplace or a public place and interview him or her on any of the topics listed above.

Learners are asked to discuss in small groups or in the group at large what was learned from this exercise in comparison to the previous one. Usually, learners find interviewing a stranger easier than interviewing a classmate, neighbor, or colleague. The use of a tape recorder is new to learners at first, but it becomes second nature once they get over the novelty. In the artificial constraint of a class time period, I ask students to find another student on campus at a snack bar, the Starbucks, or anywhere available and interview the student for 15 minutes.

Prompt:　What is your view on the quality of education you are receiving here as you work toward your degree?

Be sure to introduce yourself and ask permission to tape the individual. Explain you are taking a class where you need to practice interviewing and how helpful it would be if the person would agree to be interviewed. Get some basic information from the student such as age, how many years at the school, what degree program, why he or she selected this university and this campus, and general descriptive information before beginning the interview. Probe with follow-up questions as needed.

When class members return to discuss this and what they learned from interviewing a stranger, they often remark on how willing people are to talk about themselves. At first, there are awkward feelings talking to a complete stranger, yet in the safety of the university community, people open up. They disclose quite a bit of data and even go into specific classes and experiences that have markedly changed them. Usually, people are ready and willing to be interviewed. That is one thing you can count on in the social word, the fact that people love to talk about their work and their lives. When was the last time you had the opportunity to talk about your work with someone? It doesn't happen very often.

Discussion Points Following This Exercise:

1. How did you select and approach this person, and what did you learn from this?
2. What would you do differently if you repeated this exercise?
3. Rate yourself in terms of the kinds of questions you asked the interviewee.
4. What advice would you give others trying this exercise?

Now, take time to enter into your reflective journal your thoughts on the interview process and how you rate yourself as an interviewer in training.

Exercise 4.3

The Oral History Interview ■

The oral history interview or series of interviews describes the story of one person's life or a collection of individual stories told together. For example, the oral history of one firefighter who was at ground zero on September 11 could stand alone as an individual oral history. A collection of oral histories of multiple firefighters who lived through that experience would be a collective oral history project. The beauty of the oral history interview is that it allows us to capture the memories of people who have had any number of experiences. The meaning made of the storytelling and what we learn from the stories are critical to understanding a given period of time or a specific event or series of events. Oral history is often a vehicle for the outsiders and the forgotten to tell their stories. Currently, as I write this text, many researchers are on the way to Haiti to document the events unfolding there by interviewing survivors of the earthquake. Likewise, we are now seeing many oral histories in print from the various oral histories of Hurricane Katrina. In oral history, we tell the story through spoken text, written text, video text, or all of these media.

Oral history grew out of the oral tradition. Formal written work about oral history has emerged in the last century. Since then, we have experienced many evolutionary stages in the development of the field. Today, we are in the center of a monumental stage, that of the digital movement. This enables us to move forward experimenting with new lenses and technologies.

For this exercise, find two members in your family that represent two generations to interview about their memories of growing up, for example, your mother and grandmother, your father and grandfather, your mother and sister, your mother and brother, and so on. Use a video camera or digital voice recorder.

Goal: To find out what was experienced as a child growing up including going to school, friends, hobbies, early goals in life, key events, and so on.

Interview Protocol Sample

1. Can you remember what it was like growing up in (name place)?
2. What memories do you have from that time?
3. What do you recall from your high school days?
4. Do you recall your best friends, and what are they doing today?
5. Can you talk about your hobbies?
6. Can you talk about your dreams or goals that you wished to accomplish?

Ask the same questions of both generational members. Tape the interview, and take notes if possible. Practice transcribing the interviews. Later, write up your thoughts on what you learned about these two members of your family. Compare and contrast the generational differences and similarities. Are there key themes that emerge from the interviews? What would you like to go back and ask? If needed, do a follow-up interview.

If you do not have family members nearby to interview, try interviewing someone in your neighborhood who is a war veteran or is active in your community or is of general interest in some way. Public servants such as police officers, firefighters, elected officials, teachers, nurses, or physicians may be willing to tell you their life stories and how they came to their chosen paths in life. The point is to document their memories and interpretations of events in their lifetimes. Key oral history resources and texts will be listed at the end of this section. Digital oral histories are prolific on YouTube and on various oral history websites.

Exercise 4.4 ■

The Focus Group Interview
Demonstration Exercise ■

Demonstration: Some learners are anxious to use focus groups as a qualitative research technique, given the nature of their purposes of the study, the time line for the study, and their resources. Focus groups are not a panacea, but they do offer a way for researchers to *focus* on a topic with a given group. I ask learners to immediately read at least one text on focus groups, usually the text *Successful Focus Groups,* edited by David Morgan (1993). I use this exercise when I am fortunate enough to be working on a funded project that allows for learner participation. When on a funded project, I involve all students who volunteer to work at the actual sites. If not working on a funded project, learners may benefit from this demonstration exercise. Remember that focus groups mean just that: *focus* on one and only one topic. This exercise emerged from an actual case of a study we did of a drug-free school program in northeast Kansas. The topic is one learners and community members can talk about because everyone has something to say about drug usage in a given community. For this demonstration focus group, I ask for volunteers.

Topic: Drug availability in your school, community, and neighborhood.

Purpose: To find out how students, teachers, and parents feel about drug usage and drug education in your school.

Sample: Note that these questions may be modified for each focus group made up of parents, teachers, and students. These were the basic questions used by the trained moderator of the groups, and they were rearranged according to the composition of the group of nine members:

1. Can you talk about whether or not you feel safe in your school or community?
2. Do you have some thoughts on how your school or community is doing regarding alcohol and drug problems with students? Can you describe for me what you know about this?
3. What does your school or community do to educate you about the use of alcohol and drugs?
4. Do you think drugs or alcohol are easily available to students?
5. What kinds of rules do you have about alcohol and drugs?
6. If you ever had a problem with alcohol or drugs, who would you approach to talk about it? Can you explain your thoughts about your choice of this individual?

In this demonstration exercise, I ask for a volunteer for the moderator who must keep the group on task. Also, I give out role-taking cards with a role for each of the seven volunteers for the focus group. Each learner draws a card at random; here are some of the roles taken by participants:

Card 1: Play yourself.
Card 2: Play yourself.
Card 3: Play disagreeable, and refuse to answer every other question.
Card 4: Play overly agreeable, and agree with everyone.
Card 5: Say as little as possible, and only speak when asked a direct question by the moderator.
Card 6: Say nothing regarding drugs, and talk about every and any other subject.

(Continued)

(Continued)

Card 7: Try everything you can to get out of answering any question, and keep asking to leave the group.

These seven types can be found from time to time in any given focus group in the real world, which is why I have designed these roles partially to train moderators and partially to allow learners to see that focus groups require a great deal of patience, fortitude, tact, diplomacy, and tenacity. Some warnings about focus groups include being ready to deal with people who cry in the group, people who may lose their temper or composure, people who argue, and so on. If a moderator finds a situation where things get problematic, he or she can stop the group to take a time-out and then make a decision about continuing on or not. Thus, moderators need training in dealing with the public and must have at least the semblance of calm. Also, as the researcher, you should avoid being the moderator, for you cannot be entirely impartial. You need to find a competent moderator and train the moderator in questioning and moving the group along.

Forms, Uses, Strengths, Weaknesses: The focus group technique is one of the most common approaches to research in the behavioral and social sciences. A focus group is a group interview with a trained moderator, a specific set of questions, and a disciplined approach to studying ideas in a group context. The data from the focus group consist of the typed transcript of the group interaction.

Uses for and Ways to View Focus Groups

Self-Contained Research Technique: Supplemental technique for qualitative and quantitative studies (see Morgan, 1993). While you may get some qualitative narrative data from a focus group, this does not necessarily make your study qualitative. In order to have a qualitative study, you should have a qualitative theoretical frame to guide the study and also good qualitative questions. Most often the focus group is used to augment a survey.

Focus Groups Are Useful For:

1. Orienting oneself to a new field
2. Getting interpretations of results from earlier studies

3. Exposing professionals to the language and culture of a target group (i.e., bridging the gap between the professional and the real-world target group, as with the use of focus groups in medical research)

Strengths:

1. The major strength of focus groups is the use of group interaction to produce data that would not be as easily accessible without the group interaction on a specific topic. This can also be flipped into a weakness, as it is not contextualized. In other words, the individuals in the group are not interviewed separately to find out their own life stories or the lived experiences that relate to the specified topic of the focus group.

2. Participants' interaction among themselves replaces the interaction with the interviewer, possibly leading to a greater understanding of participants' points of view.

Weaknesses and Disadvantages:

1. Focus groups are fundamentally unnatural social settings when compared to participant observation and interviewing selected participants over time. They are also almost totally decontextualized. In other words, you focus on a given topic without getting all the life stories of the group members and all the related stories to the topic. If, as a qualitative researcher, you are trying to capture contextualized data, you may need to reconsider this choice of technique.

2. Focus groups are often limited to verbal behavior and again only on a specified particular topic. There is no room for additional information to be collected. There is limited opportunity to check back and verify on-site if the behavior actually matches what is stated.

(Continued)

(Continued)

3. Focus groups depend on a skilled moderator, who is not always available when needed. The principal investigator of a research project cannot ethically act also as the moderator. The time and money involved in training a skilled moderator is often a major problem when planning a focus group.

4. Focus groups should not be used if the intent is something other than research (e.g., conflict resolution, consensus building, staff retreats, professional development, or attitude changes). Too often, individuals are giddy over the focus group without realizing that, in a qualitative research project, one needs to ask questions of a qualitative nature.

A Sample Approach to Focus Groups:

1. Have a planning phase: Allow for 4 to 6 weeks. *Identify your goals.* State the purpose of the focus group precisely, in one sentence, if possible. Then, identify the ideal membership of the group.

2. Limit the number of groups to perhaps 2 per week over 6 weeks. Members and moderators burn out easily. The logistics of planning the focus group, picking up participants, if need be, and scheduling a site are often overwhelming. Thus, we've found that no more than two groups per week allow us to be true to our study.

3. Transcripts: Plan for 8 weeks or more. Creating transcripts is a difficult and time-consuming task, and we often underestimated the time we needed for completion. Through experience with the drug-free schools case, we learned to allow 3 to 6 months per project, with all conditions being favorable and only a few people backing out of the focus group.

After you identify your goals, I recommend applying these rules of thumb, which were constructed after polling our participants in the project:

Dilemmas: Again, these dilemmas arose from the actual case:

1. Who brings up controversial issues, the moderator or the group?
2. How far does informed consent go if unusually personal information is revealed?

In working for specificity of, depth in, and understanding of the social context in a given study, the intent of the moderator is always to *get the story.* We found that a skilled moderator is essentially the key to a successful focus group.

Some Myths About Focus Groups:

1. Focus groups are quick and cheap.
2. People will not talk about sensitive issues in groups.
3. Focus groups must be validated by other methods.

The following checklist was developed and modified from experience and suggestions from the Morgan (1993) text for the purpose of assisting learners in this process.

Advance Notice:

1. Contact participants by phone 1 to 2 weeks before the session.
2. Send each participant a letter of invitation.
3. Give a reminder phone call prior to the session.
4. Slightly overrecruit the number of participants. There are always last-minute dropouts.

Questions:

1. The introductory question should be answered quickly and should not identify status.
2. Questions should flow in a logical sequence.
3. Probing questions should be used as needed.

(Continued)

(Continued)

4. *Think back* questions (questions that begin with the words, "Think back," for example, "Think back to when you were teaching your first class . . .") should be used as needed.

Logistics:

1. The room should be comfortable, quiet, and satisfactory in size.
2. The moderator should arrive early with name tags for everyone.
3. Have appropriate digital equipment.
4. Experts and loud participants should be seated near the moderator.
5. Shy and quiet participants should be seated across from the moderator.
6. When having a meal, limit selection and get down to business.
7. Bring enough copies of handouts, visual aids, storyboards, and so on. Always have extras on hand.

Moderator Skills:

1. Be well rested, alert, and prepared, and practice your introductory remarks beforehand so as to deliver them smoothly. Remember questions without referring to notes.
2. Avoid head nodding and comments leading the members to a particular conclusion.

Immediately After the Session:

1. Prepare a brief summary of key points. Check to see that the recorder captured everything.
2. Get the tapes to your transcriber, or begin transcribing immediately.
3. Check your field notes as a checks-and-balances tool, and keep a list of all participants' names, addresses, and phone numbers.
4. Prepare your report with excerpts from the transcripts in the body of the report.

Exercise 4.5   ■

Presenting Interview Data as a Found Data Poem   ■

Example: Found Data Poetry Assignment

Found data poems are essentially *found* in the text of interview transcripts, documents, or even spoken words. For the purposes of this assignment, select a portion of an interview transcript or written text of any kind to practice creating an original poem from the words in front of you. You invent the themes and content based on the words in the transcript or other written text. It is best to keep the words of the interviewee or author as much as possible. Feel free to concentrate on meaning and interpretation.

Create a Found Data Poem of Your Choice:

1. Find at least three lines.
2. Use at least one metaphor or symbol.
3. Develop a clear theme or point of view.
4. Use at least one image that captures the meaning of the text.

Be sure to reference the source of the poem, for example, Transcript A. Include a sample of the text the poem is based upon. You are tapping into the right side of your brain, an often underused portion. We will share this with each other in class in small groups and with the group as a whole.

Sample Excerpt From an Interview of a Female Assistant Superintendent in Response to the Instruction, "Describe a Typical Day."

This is a part of an hour-long interview where the speaker shared her educational background, her goal of helping kids, and her caring for their well-being while striving for excellence. This was an interview I conducted recently in a project on female leaders.

(Continued)

(Continued)

A: Well, it's been an atypical day so. . . . And that's something
I've been thinking about. There isn't. . . . There are typical
days, and they're boring. The typical days are the days when
you're sitting and working on paperwork for the state, and
working on budgets, and trying to analyze test scores to make
them meaningful to the teachers and to the . . . and
whatever. So, those are the typical, boring days. This is our
second week of school, so there's no typical beginning of the
school year. Now, I'm spending more time supporting
teachers, right now new staff. Right now, I'm doing . . . pulling
on my special ed background. I have a little guy who is in one
of our self-contained classrooms, but he's struggling with the
transition coming back to school, and mornings aren't good
for him, and he's got a new teacher. And the principal in that
school is on maternity leave. And the principal who is filling in
was a little panicked. And so, we met and talked about
strategies for this little guy that, no, you know, in first grade,
he's not ready for therapeutic day school. He's not hurt
anybody. Everything's fine. It will be okay. We have a
controversy going on right now related to curriculum materials
that have been selected for students' optional use, optional
reading. So, we've been laughing and . . . on one
hand . . . and cringing on the other because we're responding
to one parent's concern.

Atypically, we have only heard from one parent who has a
concern about a book that was on the summer reading list. Kids
take home a list of six or seven books that are optional. The kids
give a synopsis of the book at school. They talk about 'em. And
if you don't like any of those books, you can read any other
book in the whole wide world to choose from. And this one is as
much young adult literature as it is controversial themes
because it gives us the opportunity to support kids as they worry
about these things and. . . .

Q: Can you tell the name of the book?

A: It's *Fat Kid Rules the World* by Kale Going. And the themes really
are friendship, not giving up, perseverance. A student in there

contemplates suicide. He's had a very tough time. His mom's died from cancer. His dad's an alcoholic. He's in an abusive home situation. And he is befriended by a homeless teen, who is a gifted guitarist, who asks this kid to join his band and play the drums. And, it basically is about acceptance, and you know it's a great story of redemption. It's a wonderful story. And the parent who objects is objecting based on the proliferation of the F-word. And, it is in there and it . . . kids are in Brooklyn. And, interestingly enough, but it's not really spoken out loud. It's in this kid's thoughts. That she's objecting to the normal, sexual fantasies of teenagers. He's describing a person and saying, "No, not this one, not the one with the large breasts, you know, the other one" . . . physical features. So, you know, things like that. This parent has, you know, not accepted that the fact that her child was not required to read the book and . . . she did not ask for the book to be banned from the library. I think she just asked for it to come off the summer reading list. However, that has snowballed to some right-wing websites . . . Concerned Women for America, the Illinois Family Network. I don't know which all . . . SaveLibraries.org. And, we have been getting interesting e-mails from basically all over the country and Canada.

Q: Like what kind of interesting e-mails?

A: Oh, some that are saying. . . . One was, you know, "If I knew where Osama Bin Laden was, I would turn him in, but first, I would tell him where your school was so that he could bomb it, hopefully, when there were no children . . . on the weekends when no children were present."

You know, "You're responsible for the moral degradation of children and the increase in rapes and murders and school shootings because children have read . . . because we have forced children to read this book." Personally, I, as a woman, I'm not fit to lead, even though Lord knows how that's connected. I'm not sure. We were joking that a bunch of us are Catholic, so we were all damned already anyway, so it didn't matter, so you know. . . . The board's been very supportive. They. . . . You know, they listened to 1 hour and 10 minutes of different people

(Continued)

(Continued)

expressing their viewpoints. Some people—I'd heard it on the radio—a board member from a distant suburb came to express her concerns. Somebody from our library board came. The phone call came from one of the mayors in one of the towns that we serve. But, the board was supportive and said that, you know, no, this was not required reading, that the teachers did inform students of that. We did make an error in not informing parents that they might want to be alerted to this fact. And so, we are looking at our selection policy and how we let parents know where to write . . . if they want to follow up. But in all of this time since our last board meeting, and since this started in mid August, we have only heard from that one parent.

Q: Have you been threatened, or has anybody?

A: I have not personally been threatened. The junior high principal has been threatened. You know, "When someone comes and murders your family, it'll be because of how you taught them." Rather interesting. No one from the immediate community. . . . No other parents in the community. . . . There's an article in today's paper; we had a prepared statement to share with people who called, anticipating. . . . And this is the daily paper. This is not like the local Podunk paper. We had one parent phone call with a question or concern. And we did end up sending a note home today, you know, saying that you know we didn't believe that the threat was really credible but that we did have, you know, that there was a police presence.

. . . It (this book) was on the book list for incoming eighth graders, so that would be 13- and 14-year-olds. They talk about these things. And, some statistic that I had recently came across said three to five. . . . Three out of five teenagers contemplate suicide at one point. So, um . . . yeah, it's kind of important to maybe say, yeah, there's a place to talk about this. It was . . . It's probably toward the young end of the age spectrum that the book might be appropriate for. And, we did have a parent come and talk in support of the book. Her student had read it during seventh grade. He's a very capable student. And, as a parent, she also read the book and thought it was a perfect avenue to discuss some of these difficult situations, which some of our students live in. You know we don't. . . . Not everyone lives in a two-parent home where they go to church every Sunday and their other social, emotional, and physical needs are met. Literature is one way to talk about kids who are sitting next to you or are on the other side of the country who don't have those same experiences. This student obviously and this parent, who was able to sit and talk with her child about these things, who a year later recalled that the themes were, you know. . . . Didn't recall the swearing, didn't recall the sexual fantasies, recalled the overarching themes. We have students who haven't read an entire book probably since they were in second grade, who were talking about how fabulous the book was, kids whose lives unfortunately probably mirror the protagonist's life. . . .

And, it is a big topic for kids. And, it is an issue that we think needs to be explored. And, it was optional book. It's for a lit circle about self-acceptance. But that maybe there's another . . . a better option out there. Because just because the kids can read at that level, it doesn't mean that maturity-wise they're able to.

Q: Now, thank you for sharing the e-mail that was sent to you, but for the purposes of the transcript here, can you describe the contents briefly of the e-mail that you received? This is from, again, only one parent over this one book.

A: Well, this is a person. I don't even know who it is. He equated the ability of a woman to be president and whether or not he should vote for a woman as president with my role as an assistant superintendent in this school district and the analogy of someone who comes upon a boat that has a little bit of water in the bottom and drills a hole in the boat to let the water out and continues to drill additional holes to let more water out, as of course more water's rushing in. And some man miraculously comes and says, "No, no, maybe we should stop drilling holes and plug the holes instead," and saves the day. And, who should we vote for, the man who plugs the holes or the woman who has drilled the holes? And, in his warped sense

(Continued)

(Continued)

of the world, he also sent. . . . Prior to my e-mail, my superintendent received an e-mail with just a shortened version of that analogy but accusing him of being a hole driller as well. But, I obviously am not . . . or women in general are not fit to be president because we obviously must be the hole drillers. There might be some warped men who also are stupid enough to drill holes with us. But, we as a gender probably are sucking them into it. So, that would be atypical. . . . If I get an e-mail from a parent, it's generally related to a curricular concern or a testing question.

Q: So, and let's say you get this e-mail then. How do you deal with it? What strategies do you use, or what is that like?

A: Well, I'm not deleting it, only because I'm saving it just for posterity. Other than that, we get a lot of good laughs out of it. We shared it around the office. And you go on with your day. Exactly. It will be good for a laugh at many times in the future.

(End of excerpt)

I shared this small excerpt of a larger transcript with my class. Take a look now at the poetry created by Jill Flansburg upon reading this excerpt. She decided to do four haiku. Haiku are generally 17 syllable Japanese poems done in 3 lines.

Poems

By Jill Flansburg

Parents misconstrue
The teachers, the kids, the book,
Narrow-mindedness.

Poise under pressure
Never a typical day
But I really care.

I'm damned if I do and

Damned if I don't
Too vital to quit.

Parents find fault
Intolerant of teachers
And the kids miss out.

Now, try your own poem based on the transcript above. If it is possible to work in a group, try this as a group project.

Another example of found data poems can be seen here. I looked at the website for the electronic journal *The Qualitative Report*. There was a celebration of the journal's 20th anniversary in January 2010, and there was a call for testimonials, poetry, and videos to be sent to the celebratory conference. Here is some poetry I created for that celebration.

An Example of a Haiku: Evocative of the 20th Anniversary of *The Qualitative Report*

Meaning

By Valerie J. Janesick

Meaning comes with solitude
Like the egret on the water
Silence again.
(December 16, 2009)

Next, in order to create a poem, I went to the website for *The Qualitative Report* and used the editorial statement as the data for this poem.

We Are *The Qualitative Report*

By Valerie J. Janesick

We are
Peer reviewed with
Passionate ideas, and
Critical collaboration,
Debates, and topics, and

Research possibilities
Based upon the process.
The hallmark is assisting,
Not rejection rates
Manuscript as centerpiece in the
Development Process,
Assigned a team to
Weave collective comments.
We grow as authors and mentors
We are a learning community
We give back
We benefit
We trust
We are.

(December 16, 2009)

After the experience of interviewing, creating data poems, and creating codes and categories, learners appreciate the opportunity to practice, individually and in groups, the demanding task of analyzing interview data. Within the parameters of the class, members have already taped an interview on views of friendship. They transcribe 10 minutes of tape and then begin working in groups of three or four to find major categories from the data. This is a seemingly small task, yet it takes about an hour of class time. Members go through each other's transcripts and field notes, and they listen to the tape as often as they need to. Each group comes up with a set of major and minor categories. Remarks after this exercise often include the following:

I was amazed at how the categories popped out of the data.

This was harder than I thought because I forgot to take field notes during the interview, and so I really needed to know the nonverbals.

This was much easier when I had someone with whom to check my categories. Now, I know why having an outside reader of field notes and transcripts, as suggested, is a good idea.

Figure 4.1   "Major Themes Emerging From Observations and
Interviews" by Carolyn Stevenson

William Wallace College		Jane B. College
Developmental steps for integrating technology into the curriculum	⇔	Developmental steps for integrating technology into the department and curriculum
Change involved motivation and collaboration	⇔	Motivating others and teamwork needed for change
Leadership needed for change	⇔	Leadership as change agent
Continued growth seen in the future of technology	⇔	Accessibility
Faculty resistance		Underprepared students
Use of technology in the curriculum		
Distance learning is not for every institution	⇔	Distance learning is not for every student

This teaches me to have another interview and get more data.

The person I interviewed taught me better questions to ask.

As I looked over the transcripts, I realized how much I had already forgotten about the focus group. Now I know I have to transcribe everything because my memory is not what it used to be.

Through the experience of working on transcripts in the small group, members feel more confident in dealing with their miniprojects as far as analysis of data. Learners look for major themes, key words, and indexes of behavior and belief, and they make an initial list of major and minor categories. Every attempt is made to look for critical incidents, points of tension and conflict, and contradictions to help in the purposes of study. In the class situation, most students found working with a group or a partner to be helpful and illuminating.

Figure 4.1 shows an example of developing a model or visual representation from your study after poring through the interview data. This model is from Carolyn Stevenson.

Exercise 4.6

The Digital, Virtual Interview and Google Groups

The amazing growth in the number of articles, books, websites, LISTSERVs, and interviews on YouTube are indicators of the interest and growth of Internet inquiry. In terms of qualitative interviewing, the Internet affords many opportunities for interviewing without the boundaries of traveling or being physically present for an interview. On the other hand, a number of issues arise from examining data posted on the Internet. Despite the relocation to the virtual world in digital form, Internet inquiry still contains the basics of qualitative research. You will still be seeking information about lived experience and the social context. You will still need to complete your transcripts of any interview. In this exercise, there are two parts:

Part One

Try an interview with someone you know on Skype. To use Skype, simply go to www.skype.com to create your free account and have access to video and voice capability on their system. Start with 10 minutes, and see how this feels to you. Use the questions listed earlier, such as, "What does your work mean to you?" Save the interview. Or if this is not possible, view a completed interview on YouTube. Find three major themes in the interview to discuss with a critical friend or partner in class. Write about what you learned in your reflective journal.

Part Two

Now, go to the site groups.google.com (formerly Deja News), which is part of a cluster of groups on Google. Here, you may start a discussion about your interviewing techniques and talk about what you learned from personal online interviewing or pulling three themes out of an extant YouTube interview. You can do these things online with Google Groups.

1. Create a group.
2. Set up your group.
3. Invite people to join it.
4. Discuss online or through e-mail.

You may discuss online or in e-mail, and you can create a custom page for this. After you select your fellow group members, ask each to talk about the experience of interviewing someone online. What do you see as the similarities and differences in your real-time, face-to-face interviews?

Becoming comfortable is a must in developing your interviewing and writing habits. It is not as awkward as you might imagine. Also, visit YouTube for interviews already uploaded online. Here, you will see examples of oral history interviews and face-to-face interviews on various topics in multiple areas of study, such as gerontology, family health, nursing, education, business, and sociology to name a few.

Exercise 4.7

Interviewing Someone Twice

Goal: Interview one person twice, so that you have the experience of going back for an interview. Interviews should be approximately 1 hour in length each. Be sure to allow time between interviews. Do the first transcription then go on to the next interview.

What Does Your Work Mean to You?

Select an educator or other professional to interview about what work means to that person. The first interview should have some basic, grand-tour questions.

Interview Protocol A: First Interview

1. What does your work mean to you?

2. Talk about a typical day at work. What does it look like?

3. Tell me what you like about your job.

4. Tell me what you dislike about your job.

(Continued)

(Continued)

5. Where do you see yourself in 5 to 10 years?

6. Is there anything else you wish to tell me at this time?

7. Is there any question you have about this interview project?

You create the questions for Interview Protocol B.

Create the questions based on what you found in the first interview and to get to the goal of the interview. This should take at least 1 hour but no longer than 90 minutes per interview. Remember that people are busy, and they most likely can only give you an hour at a time. Be sure to have your original field notes so you can probe into areas of the first interview during the second interview. Be sure to tape all this. Be sure to get informed consent either by signature or on the audio tape.

Write a report on the interview that includes at least the following items:

a. Describe in detail why you selected this person. Add a photo if needed.

b. Provide a list of all the questions asked in each interview and label them Interview Protocol A and Interview Protocol B.

c. Summarize the responses from both interviews in some meaningful way with precise quotations from the interview.

d. Pull out at least three themes from the interviews.

e. Tell the story of what this person's life work means to this person.

f. Include the signed consent form or a form stating that the interviewee gave consent on the digital tape, and you sign that form, too.

g. Discuss any ethical issues that may have come up.

h. A sample of 7 to 10 pages of your best transcript from the tape should be added as an appendix.

i. Include two or more pages of your own reflections on your skills as an interviewer, as a researcher, and what you learned from this project; be sure to mention any difficulties that came up and what you would change the next time you conduct an interview.

j. Be sure to use a minimum of five references from this term's texts.

k. Be sure to create a title that captures your themes.

Remember that this is a narrative research paradigm, so you should write this in narrative form as if you were telling this person's story. The story is about the person's life work.

Exercise 4.8 ■

Practicing an Online Interview ■

Goal: To conduct an online interview using Skype.

Because the purpose of conducting an interview is to get data, that is, to get some information from a participant about a certain topic, students born and raised in the digital era are besieging me with requests about digital interviewing. This semester, two students are trying interviews online. Because Skype is free online, open an account, and you will have video and audio capability for an interview. You may save the interview on your computer for future transcription. Use Exercise 4.7 as your guide to conduct two interviews online. In addition to the information listed there, add in your report a description of why you elected to do an interview online, what your view is about interviewing online, what the positive and negative characteristics of the interview were, and any suggestions for those considering online interviewing.

■ Summary of Chapter 4

In most cases, interviewing is like a duet or *pas de deux* in dance. Two people are communicating with one another and, ideally at least, understand each other whatever the context. The major pitfall in interviewing is not being prepared with mechanical materials, questions, and good communication skills. Direct experience, practice, and reflection are the strongest assets of the interviewer and the dancer. As themes and categories jump out at you from the transcripts and field notes, learn to develop models of what occurred. Models are your visual representations of what occurred and your interpretation of what occurred in the study. Practice writing in your researcher reflective journal every day to have a road map of your thinking about inquiry and interviewing.

■ Resources to Improve Your Knowledge of Interviewing Skills

Berg, B. (2007). *Qualitative research methods for the social sciences* (6th ed.). Boston: Allyn & Bacon.

Denzin, N. K. (1989). *Interpretive biography.* Thousand Oaks, CA: Sage.

Eisner, E. W., & Peshkin, A. (Eds.). (1990). *Qualitative inquiry in education: The continuing debate.* New York: Teachers College Press.

Janesick, V. J. (2010). *Oral history for the qualitative researcher: Choreographing the story.* New York: Guilford Press.

Kvale, S., & Brinkmann, S. (2009). *Interviews: An introduction to qualitative research interviewing.* Thousand Oaks, CA: Sage.

Lortie, D. (2002). *Schoolteacher: A sociological study* (2nd ed.). Chicago: University of Chicago Press.

Marshall, C., & Rossman, G. B. (1989). *Designing qualitative research.* Thousand Oaks, CA: Sage.

McCracken, G. (1989). *The long interview.* Thousand Oaks, CA: Sage.

Merriam , S. B. (1997). *Qualitative research and case study applications in education* (Rev. ed.). San Francisco: Jossey-Bass.

Mishler, E. G. (1986). *Research interviewing: Context and narrative.* Cambridge, MA: Harvard University Press.

Morgan, D. (1988). *Focus groups as qualitative research.* Thousand Oaks, CA: Sage.

Morgan, D. (Ed.). (1993). *Successful focus groups.* Thousand Oaks, CA: Sage.

Patton, M. Q. (1990). *Qualitative evaluation and research methods.* Thousand Oaks, CA: Sage.

Reinharz, S. (1992). *Feminist methods in social science research.* New York: Oxford University Press.

Roulston, K. (2010). *Reflective interviewing: A guide to theory and practice.* Thousand Oaks, CA: Sage.

Rubin, H. J., & Rubin, I. S. (2005). *Qualitative interviewing: The art of hearing data* (2nd ed.). Thousand Oaks, CA: Sage.

Salmons, J. (2010). *Online interviews in real time.* Thousand Oaks, CA: Sage.

Witherell, C., & Noddings, N. (Eds.). (1991). *Stories lives tell: Narrative and dialogue in education.* New York: Teachers College Press.

CHAPTER 5

The Creative Habit ■

Being creative is not a once-in-a-while sort of thing. Being creative is an everyday thing, a job with its own routines. That's why writers, for example, like to establish a routine for themselves. The most productive ones get started early in the morning when the phones aren't ringing and their minds are rested and not yet polluted by other people's words. They might set a goal—1,500 words or stay at their desk until noon—but the real secret is that they do this every day. They do not waver. After a while, it becomes a habit.

Twyla Tharp (2003), Choreographer
The Creative Habit

In the field of dance, creativity is valued and practiced on a regular basis, thus a habit is formed. In qualitative research, it is also helpful to focus on this habit and practice creativity. In this section are exercises I have adapted for allowing learners to develop techniques in interpretation and for defining the role of the researcher. The practice of narrative techniques can only assist the researcher in training. Modern dance, as an art form, is characterized by a language of movement, and no one speaker makes quite the same statement. Likewise, the dancer explores new ways of moving, which include creative experiences and interpretation. Similarly, in yoga, as the practitioner begins, he or she needs to explore the postures in order to reorganize the spine. In yoga, too, one must acknowledge all the inner systems of the body in order to use the mind more fully. In working with mostly graduate students in education and human services, who have spent many years in bureaucratic settings, these are the exercises that cause the greatest disequilibrium. At the same time, they offer the practitioner the most opportunity for self-awareness. In fact, many individuals came to the realization that qualitative methods were much too demanding for them, and they would prefer to work in another paradigm. This is, of course, a creative awakening in itself. Just as everyone in dance must end up asking the question, "do I really want to be a dancer?" the prospective qualitative researcher must ask the question, "do I want to be a qualitative researcher?" Nevertheless, as a teacher, I see these exercises as a beginning point for self-identification and, consequently, valuable tools for any researcher. For the prospective qualitative researcher, these exercises help to instill an awareness of the importance of the role of the researcher. In each qualitative research project, the researcher must explain fully the role of the researcher. I require potential qualitative researchers to be able to describe and explain their own roles in their individual projects. We begin with a seemingly simple exercise, again modified from my days as a student of drawing. The exercises in this section are framed within John Dewey's notion of the development of habits of mind and the aesthetic as part of everyday experience. In Dewey's time, there was a prevailing modernistic dualism that separated the aesthetic from the world of ordinary experience. In this postmodern time, I have constructed these exercises to help the individual address this dualism and engage in ordinary experience as aesthetic experience. Be sure to write in your researcher reflective journal, describing what you did, how you completed these exercises, visual images of the exercises, and what you learned as a writer.

■ The Role of the Researcher and the Researcher Reflective Journal

The following series of exercises are focused on developing your role as a researcher in order to understand yourself in deeper ways of knowing. In order to do that more fully, I emphasize the importance of the researcher reflective journal. By putting journal writing into a historical context and understanding that there are multiple types of journal writing, the learner may be more definitive in understanding the self and the researcher as a research instrument. In conducting a dissertation study or any research, the value of keeping a substantive journal is that it may be used as a data set to complement other techniques. We begin with the most personal of exercises, that of writing your name in new ways.

Exercise 5.1   ■

Variations on Writing Your Name   ■

Purpose: To write your name as many times and in as many
ways as possible on a sheet of 8½" × 11" paper.

1. First, write your name with each hand—left hand
   first, then the right hand.
2. Next, write your name upside down.
3. Then, write your name diagonally, in the shape of an X.
4. Now, write your name from right to left, as many
   cultures do.
5. Now, write your name as you usually do.
6. Now, close your eyes and write your name.

Discuss and compare your reactions to writing your name in
these varied formats. Pair up with someone in class and view each
other's writing. Discuss and share your reactions.

Problem: To liberate yourself from the usual writing of your name.
To jog your multiple intelligences. To stimulate the right
side of your brain.

Note: I recently took a drawing class, and the teacher used
some of these variations. This reminded me of Betty
Edwards's (1986) *Drawing on the Artist Within*, which
was a follow-up to her renowned earlier works, *Drawing
on the Right Side of the Brain* (1979) and *The New
Drawing on the Right Side of the Brain* (1999). I
encourage all my students to read these texts or a
reasonable equivalent.

Time: Take as much time as needed.

(Continued)

(Continued)

Activity: Create a visual representation of one's own name.

Aim: The purpose here is to give the person the opportunity to break away from the typical expression of writing his or her name. This is so personal an exercise and produces such confusion for some because it is the first time in a long time that they have put creativity into action. I also see this as the beginning step, a first step, in recognizing the active nature of the role of the researcher. The researcher is not passive. The prospective researcher must begin to recognize his or her own investment in the research project and how critical the definition of his or her role in the project remains. This exercise generates a great deal of discussion and makes the individual learner respond to something often overlooked—the way we write our names. Overall, the goal is to sharpen awareness. Notice how you, as a learner, approach each of these phases of writing your name. Which one was most difficult?

Discussion:
    1. How did you approach this exercise?
    2. What was most difficult for you in each of these phases of name writing?
    3. What have you learned from this?
    4. What have you learned thus far about your skill as an observer? A researcher?

Evaluation: Continue working on self-evaluation for your researcher reflective journal and portfolio.

In this second exercise, the learner progresses from the internal to the external arena by using a camera to document a familiar social setting.

Exercise 5.2   ■

The Camera as an Extension of the Eye, the Eye as an Extension of the Soul   ■

Purpose: Take photographs of any area of campus or your workplace over 1 hour.

Problem: To document some portions of a familiar setting.

Materials: Individuals need a camera. Those who do not have a camera need to purchase a disposable camera, available for under $10.

Activity: Document as many different aspects of the environment as possible with your photos. Select your five best to share with the group. If working outside a class situation, find someone with whom to discuss your work.

Of all the exercises I use with learners, this one seems to inspire the most confidence, an awareness of one's limitations, and the most enthusiasm for finding out what kind of qualitative researcher one might become.

Evaluation: Continue on self-evaluation for your journal and portfolio.

Exercise 5.3 ■

Building a Collage:
My Role as a Researcher ■

Purpose: To design and construct a collage that represents your role as a researcher in the project you are developing for study.

Problem: To capture your perspective on your role as accurately as possible.

Time: 2 to 3 weeks.

Activity: Construct a collage on a piece of poster board that is a manageable size for display and discussion in class. A suggested size is 24 inches by 36 inches. Use any media you wish—printed text, photographs, magazine ads, newspaper headlines, objects, and so on.

Discussion:

1. How did you approach this activity?
2. What issues and ideas about your role as a researcher are emerging as you construct your collage?
3. What was the most challenging part of the activity for you?

Evaluation: Continue on your self-evaluation and overall evaluation for your portfolio.

Rationale: Students who select this activity become actively involved in representing their own feelings and ideas about their roles as researchers.

Exercise 5.4 ■

Constructing a YaYa Box or Making a Quilt Patch ■

Purpose: To design and construct a YaYa box. This is adapted from the field of art therapy. A YaYa box is designed to represent a person's innermost self on the inside of the box and the outward self on the outside. If you would rather make a quilt patch, the patch will represent some part of your inner self as a researcher.

Problem: To capture yourself as you are now in terms of your current role in your research project.

Time: Take as many weeks as you need to develop, create, and construct this, with the presentation of the box at the last class meeting.

Activity: Find a box of any manageable size, from a cigar box to a steamer trunk. Use multimedia to build your box. The inside of the box will depict your innermost feelings, thoughts, and beliefs about who you are as you participate in your research project. The outside of the box will represent your outer self or how your participants see you. Use any objects, text, decorations, and so on that you want to convey your idea of your role as the researcher. If you select a quilt patch, use a 12-inch-by-12-inch patch with any materials of your choice.

In less than two pages, describe the contents, decorations, and meaning of your YaYa box or quilt patch to accompany the finished artwork.

Discussion:
1. How did you approach this project?
2. What issues about the role of the researcher confronted you as you began and implemented this project?
3. What was the most difficult part of this activity for you?

Evaluation: Continue with self-evaluation and overall evaluation for your portfolio.

Rationale: Individuals become intensely absorbed with this activity and focus on deconstructing their own roles in their research projects. The ability of learners to go deeply into reflection on their roles and their effects on research projects is evident.

These activities lead to the next, most soul-searching of activities, that of writing the researcher reflective journal.

■ The Qualitative Researcher as User of All of One's Senses, Including the Intuitive Sense

One of the amazing strengths of the qualitative researcher, as I have written previously (Janesick, 2001), is the ability to use all the senses to undertake the research act. Sight, hearing, touch, smell, and taste often must be used to collect data. After living in the field with participants over time, the researcher also uses intuition—informed hunches, if you prefer—to plan the mode of inquiry, undertake the inquiry, and develop a way of "seeing" what is evident in the social setting. The role of the qualitative researcher demands total involvement and commitment in a way that requires, much like the artist or dancer, a total immersion of the senses in the experience. Like Dewey advises, art is the bridge between the experience of individuals and the community. So, too, the qualitative researcher is someone who must establish a bridge as a part of the community under study. The qualitative researcher takes on the implicit task of working in a given community and does not have the luxury of being distant, apart from the experience under study, or "objective." I only wish to point out that the role of the qualitative researcher is a role that embraces subjectivity in the sense that the researcher is aware of his or her own self, in tune with his or her senses, and fully conscious of what is taking place in the research project. Subjectivity is something to be acknowledged and understood. Without understanding where one is situated in the research act, it is impossible to claim consciousness and impossible to interpret one's data fully. Meaning is constructed in the ongoing social relationship between the researcher and the participants in the study. It is no longer an option to research and run. The researcher is connected to the participants in a most profound way, and that is how trust is established. This trust then allows for greater access to sources and ensures involvement from participants, which enables them to tell their respective stories. Those of us who have conducted long-term qualitative studies know that participants want their voices to be heard and do not want to be abandoned after the research project. My field, education, has a long history of researchers who come into a school, collect data, and flee. Thankfully, this is changing in terms of researchers' sensitivity to maintaining contact and a relationship with participants in their studies in order to maintain that sense of community that is part of any qualitative research project. This relationship remains as part of the research context throughout a significant period of time well beyond the end of data collection.

As I mentioned earlier, the senses are used in an intelligent way. Although sight and hearing are obvious senses employed in doing observations and interviews, the other senses may be used while conducting research at various sites. For example, the researcher may need to interview a participant at a restaurant or coffee shop. Once, while interviewing a blind and deaf research project participant, I had to sign into the person's hand, thus using touch in a way I never had before. Beyond this, however, all researchers use a sixth sense, an intuitive sense, to follow through on hunches that emerge from observing and interviewing in a particular social context. Researchers ought to have the opportunity in their training and in practice to sharpen their intuitive skills, which often open up avenues of data previously unknown or hidden. In exercises that I give my students to become better listeners and better observers, I often see the prospective researcher refine some of those intuitive skills so needed in research and life. The next two exercises are for practicing the use of all your senses.

Exercise 5.5 ■

Writing About Your Favorite Vegetable ■

Incorporating Sight, Touch, Smell, Sound, and Taste

In this exercise, select your favorite vegetable. Hold it in your hand to feel its texture. Smell this vegetable. Now write two pages describing this vegetable. Recall how it tastes, or taste it. Write another page just about its taste. Next, write another page about your favorite meal incorporating this vegetable. Now, construct a metaphor about this vegetable, and write about that for another page. Find a partner in class or at home to read your description. Ask for feedback, and rewrite your narrative on this, your favorite vegetable. Now, write your thoughts in the reflective journal about what you learned from this exercise. Can you tie this to your role as a researcher?

■ Serendipity

So, too, the qualitative researcher often stumbles onto something in the course of a research project that leads to a rich course of inquiry and was unplanned in the original design. In other words, one builds in a type of latent flexibility that enables the researcher to find, through serendipity, a tremendous amount of meaningful data for a fuller picture of the study. The qualitative researcher should expect to uncover some information through informed hunches, intuition, and serendipitous occurrences that, in turn, will lead to a richer and more powerful explanation of the setting, context, and participants in any given study. The qualitative researcher is in touch with all of his or her senses, including the intuitive sense, or informed hunches, based on key incidents and data from the research project. Furthermore, the qualitative researcher may expect the unexpected. For the qualitative researcher, the role becomes expanded in that the number of options for coming upon new data is enlarged, because one can always count on serendipity, contradictions, and surprises in everyday life, the true domain of the qualitative researcher. Furthermore, the qualitative researcher describes and explains these occurrences as part of the discussion of the research process and the researcher's role.

Analysis of data, like the dancer moving across the floor with floor exercises, consists of the actual *doing* of the work. For the researcher in progress, the researcher sifts through mounds of data; looks for emerging themes, ideas, issues, conflicts, and tension; and checks back with participants to verify the accuracy of these points in the journey. After the researcher has sifted through the data transcripts, field notes, and other documents, the good analyst uses the following guide to move on to reporting and interpreting data.

■ Intuition and Creativity in Research

> The lunatic, the lover, and the poet
> Are of imagination all compact . . .
> The poet's eye, in fine frenzy rolling,
> Doth glance from heaven to earth, from earth to heaven;
> And as imagination bodies forth
> The forms of things unknown, the poet's pen
> Turns them to shapes and gives to airy nothing
> A local habitation and a name.
>
> —*William Shakespeare*
>
> A Midsummer Night's Dream *(5.1)*

Here, I would like to discuss the nature of intuition and creativity as key components of qualitative research projects. By discussing intuition and creativity, I hope to initiate a conversation that may illuminate how we view the role of the qualitative researcher and how we may better explain that role. I will once again use the metaphor of dance (Janesick, 2000), and in this case, I see intuition and creativity as a *pas de deux*. In dance, the *pas de deux* is designed for two dancers with the idea that they move as one. They are totally connected to the final product, whatever the meaningful movement is to be and however it is to be articulated. For our purposes here, I define intuition as immediate apprehension or cognition. Intuition is a way of knowing about the world through insight and exercising one's imagination. Likewise, I define creativity in its generic sense, that is, having the sense or quality of being created rather than imitated. In other words,

I am trying to shift the conversation about qualitative research methodology and design from the linear approach to method and design to an understanding of the intuitive and the creative.

Doctoral students often discuss with me the ways in which intuition has manifested itself in their research projects. They often want to go further in exploring how they came to probe in interviews, how they decided to go back to social settings on given days, or how they revisited their interview transcripts. This is the phenomenon we seek, the act of using intuition and creativity. Historically, over the past 40 to 50 years, we have been writing and thinking a great deal about the design of qualitative research projects and about technique. Although design and technique are critical, I want to shift this conversation to go beyond technique. I would like to pause and look to writers from art, science, literature, and dance to make my key points.

I begin with the words of the Chinese master painter and teacher Lu Ch'ai, from the 1701 classic on painting, *The Tao of Painting:*

> Some get great value on method, while others pride themselves on dispensing with method. To be without method is deplorable, but to depend on method entirely is worse. You must first learn to observe the rules faithfully; afterwards modify them according to your intelligence and capacity. The end of all method is to have no method. (Chuan, 1963, p. 17)

Although Lu Ch'ai codified these remarks and the entire text in the 18th century, it is actually a formal text put together from material spanning the previous 11 centuries. The advice is relevant here to the work of the qualitative researcher. Have we not found, as we teach our classes, that learners begin with an almost slavish adherence to rules? Have we not seen, in the many methods texts, advice on how to do observations, interviews, journal writing, archival retrieval of evidence, and the like? This advice, almost prescriptive in nature to assist beginners, must be extended to include rules of thumb or information on technique, much as choreographers and stage directors do. In the case of dance, for example, mastering the rules of technique is critical but only a beginning. The dancer continues to practice those techniques daily, which eventually allows him or her to modify and interrupt movement and technique. The result is a creative act. The creative act relies on the dancer's intuition as much as physical technique, endurance, and stamina.

Likewise, the qualitative researcher may benefit from exercising creativity by being awake to the intuitive inclinations ever present in fieldwork. In thinking about and investigating what has been written about intuition and creativity, I turn now to analyzing some current writing as well as reviewing work that touches on intuition and creativity. In addition, the role of the qualitative researcher is of critical importance because the researcher is the research instrument. If we can help describe how we use our intuition and creativity in our research projects, all of us benefit. In fact, most doctoral students in the social and behavioral sciences fully explain their particular roles in their research projects in their dissertations in a section on methodology and may revisit their roles in a later chapter. For those who might be using this book outside a class setting, these basics also apply. Like the artist who uses paint and brushes or the dancer who uses movement, the qualitative researcher uses many techniques as tools to ultimately tell a story. For us, words and the power of the narrative are essential. By understanding how we use intuition and creativity, we may widen our vocabularies to understand the role of the qualitative researcher.

I want to address some of the key points in Mihaly Csikszentmihalyi's (1996) work as reported in his major text, *Creativity: Flow and the Psychology of Discovery and Invention*. Csikszentmihalyi was awarded a grant from the Spencer Foundation to study creativity as a lifelong process. In beginning the project, Csikszentmihalyi found no systematic studies of living, creative individuals aside from biographies and autobiographies. He ventured to design a 4-year interview and observation study of 91 creative individuals in the fields of literature, art, physics, and biology (although, I am sorry to say, no one from dance). Csikszentmihalyi found three ways to look at creativity:

1. The first way to approach creativity is the way we normally do in ordinary conversation. Here, we refer to those who express unusual thoughts, are interesting and stimulating, and are bright people with quick minds as brilliant. These are people with curious and original minds.

2. A second view—personal creativity—refers to people who experience the world in novel and original ways. They make important discoveries, but only they know of the discoveries.

3. The third view of creativity refers to individuals who have changed our culture in some way. For example, Michelangelo, Leonardo da Vinci, Albert Einstein, Arthur Miller, Martha Graham, Pablo Picasso, Charles Dickens, Leonard Bernstein, and Virginia Woolf would fall into this category. Viewing creativity in this way, the individual must

publicize in some form the idea that makes a shift or change in culture. Likewise, the qualitative researcher must publish or at least disseminate his or her findings from a study.

What Csikszentmihalyi (1996) found as major themes are those that qualitative researchers often discuss, describe, and explain. Creative people, he pointed out, are constantly surprised and always find new ways of looking at a given problem. He labeled their ability problem finding. I would go a step further and say that good qualitative researchers are indeed problem finders, but they are also problem posers. In any given study, a new way of looking at a setting can also be a way of posing and constructing something new. In fact, qualitative researchers are often coresearchers with the participants in a given study, and the participants open up new ways of looking at the social setting.

Csikszentmihalyi (1996) went on to say that he found that creative individuals exhibited curiosity and interest in their worlds not limited to their content expertise. They often read both outside and inside their own field. Of course, they were all content experts in various fields, such as literature, physics, biology, and music, and were curious about moving forward in their fields. In addition, they were curious about the world around them and how that related to their worlds, their fields of expertise, and their lives.

Threads of continuity from childhood to later life were another valuable finding of Csikszentmihalyi's (1996) study. Some followed convoluted and unpredictable routes to where they stood. Yet most, such as Linus Pauling, always knew they were the artist or scientist in the making. Pauling worked in his father's drugstore as a child, which sparked his interest in chemistry. Likewise, Frank Offner, the famed electrical engineer and inventor, recalled:

> I know that I always wanted to play and make things like mechanical sets. . . . When I was 6 or 7 years old, we were in New York, and I remember at the Museum of Natural History, there was a seismograph which had a stylus working across the smoked drum, and there were a couple of heavy weights, and I asked my father how it worked and he said, "I don't know." And that was the first time . . . you know, like all kids do, I thought my father knew everything. . . . So I was interested in how that worked, and I figured it out. (Csikszentmihalyi, 1996, p. 99)

Offner went on to make many discoveries. He developed transistorized measuring devices, the differential amplifier, and medical instrumentation.

He figured out how to make the measurements with an electrocardiogram, electroencephalogram, and the electromyogram. Some of his greatest inventions involved a stylus moving across a drum. So, there was a very long thread of continuity in his case.

Another example of continuity can be found in C. Vann Woodward's interest in the history of the South:

> That interest was born out of a personal experience of growing up there and feeling strongly about it, one way or another. I have always told my students, "if you are not really interested in this subject and do not feel strongly about it, don't go into it." And of course much of my writing was concerned with those controversies and struggles that were going on at the time and what their background and origins and their history were.
>
> The place I grew up was important. The environment and the time following the Civil War and Reconstruction. . . . It is the defeated who really think about a war, not the victors. (Csikszentmihalyi, 1996, p. 216)

For Woodward, again, the interest in his work began early in childhood. Likewise, Ellen Lanyon recalled her first feeling of destiny and creativity related to her grandfather's death. Her maternal grandfather came to the United States from Yorkshire, England, for the World's Columbian Exposition of 1893 in Chicago. She always believed that she would follow in her grandfather's footsteps:

> And when I was about 12 years old, my grandfather died. My father and mother put together his equipment that was left plus new tubes of paint, etcetera, and it was presented to me on my 12th birthday . . . and so I started painting. . . . I can absolutely remember the room, the place, you know, everything. I don't know what happened to the painting . . . but that's the kind of beginning that sets a pattern for a person. (Csikszentmihalyi, 1996, p. 220)

What these creative individuals show us can help us, as qualitative researchers, to dig deeply into our roles and go further in explaining the beginnings of our interest in the work we do. This can help illuminate more clearly the role of the researcher in qualitative research projects.

■ The *Pas de Deux* of Intuition and Creativity: Lessons for Today

At this point, you may be wondering about the lessons we might learn from this long-term study on intuition and creativity. Like the two dancers in a *pas de deux*, intuition and creativity seem almost as one. Like yoga, body and mind are one. They embolden our discoveries and questions, whether in art, music, literature, the sciences, or everyday life. Intuition is connected to creativity, for intuition is the seed, so to speak, of the creative act. Qualitative researchers spend a great deal of time and energy inquiring into social settings and the meanings of the actors' lives in those settings. If we take the time to carve out some space to understand the place of intuition and creativity in our work, like the dancers of the *pas de deux*, we present a more complete, holistic, and authentic study of our own roles as storytellers and artist-scientists. For qualitative researchers, the story is paramount. And nothing is so important to the story as the words we use, both intuitively and creatively. A good way to document the story is for you as the qualitative researcher to keep the story going and writing in your researcher reflective journal.

■ Writing the Researcher Reflective Journal

In working with prospective qualitative researchers, one of my goals is to inspire my students to keep a journal in order to use that as a data set in the dissertation or thesis. Along with the interview transcripts, documents from the study, and photos, now the learner adds sections of the researcher reflective journal. They read *At a Journal Workshop: Writing to Access the Power of the Unconscious and Evoke Creative Ability* by Ira Progoff (1992). This text offers an extremely sophisticated and challenging approach to deepening one's self-awareness. In my view and Progoff's, deepening self-awareness helps to sharpen one's reflections, writing, thinking, and ability to communicate. Thus, for the qualitative researcher, the meditative focus of journal writing can only help to refine the researcher as a research instrument. The ideal situation would be to work through every component of the text, which I see as a lifelong task. Because we are limited in our class to only 16 weeks together, I have adapted some of Progoff's ideas into a workable routine for my students and myself. Progoff writes about journalizing as a *life history log*. The following framework is adapted to make the student a better researcher.

Writing as Pedagogical and Research Practice:
Introducing the Researcher Reflective Journal

The reader may remember that field note writing, journal writing, and descriptive vignettes in general have been part of the exercises described in this text. I have always been struck by the power and place of writing in my career as an educator. In fact, as I wrote earlier (Janesick, 1995), most of my life consists of writing, reading other people's writing, editing, and rewriting and evaluating the writing of myself or others. What is ironic to me is that in research programs of doctoral students, so little emphasis is placed on writing as a pedagogical tool and as a preeminent focus of research dissemination. In my classes in research, students often express amazement at the amount of reading and writing required to be a good researcher, yet months or years later, they express gratitude for having that opportunity to realize that writing is a chief component of qualitative research. Earlier in this text, I introduced the notion of the *dialogue journal* based on the Progoff model. That was for the purpose of the researcher coming to an awareness of self. Now, we turn to the reflective journal during the research process to go further and come to an awareness of how your participants think, feel, and behave.

Furthermore, writing that is accompanied by reflection on that writing often leads to new questions about the research act, the study being reported, and questions in general about society, social justice, and responsibility. When learners reflect on this within the framework of their research, they often remark on a feeling of empowerment. When individuals keep journals of their own thoughts on the research process, or interactive journals with the participants in their studies, or write letters to me or other researchers, they discover and articulate their own theories about their research practices.

What results is a kind of active learning from one another so that power is decentered and the research process is demystified. In addition, writing is one of the acts of democratization of the research process. Writing engages, educates, and inspires, which can only be helpful in trying to understand what qualitative researchers do in their respective research projects.

On the Importance of Journal Writing
for the Qualitative Researcher

A journal may be used as a qualitative research technique in long-term qualitative studies. For qualitative researchers, the act of journal writing may be incorporated into the research process to provide a data set of the researcher's reflections on the research act. Participants in qualitative studies may also use journals to refine ideas, beliefs, and their own responses to the research in progress. Finally, journal writing between participants and researcher may offer the qualitative researcher yet another opportunity for triangulation of data sets at multiple levels. Journal writing has a long and reliable history in the arts and humanities, and qualitative researchers may learn a great deal from this. It is not by accident that artists, writers, musicians, dancers, therapists, physicians, poets, architects, saints, chefs, scientists, and educators use journal writing in their lives. In virtually every field, one can find exemplars who have kept detailed and lengthy journals regarding their everyday lives and their beliefs, hopes, and dreams. I see journal writing as a powerful heuristic tool and research technique and will discuss reasons for using journals within qualitative research projects in order to do the following:

1. Refine the understanding of the role of the researcher through reflection and writing, much as an artist might do

2. Refine the understanding of the responses of participants in the study, much as a physician or health care worker might do

3. Use a journal as an interactive tool of communication between the researcher and participants in the study

4. View journal writing as a type of connoisseurship by which individuals become connoisseurs of their own thinking and reflection patterns and, indeed, their own understanding of their work as qualitative researchers

The notion of a comprehensive reflective journal to address the researcher's self is critical in qualitative work because of the fact that the researcher is the research instrument. In reviewing the literature in this area, journal writing, although an ancient technique, is only now being used and talked about as a serious component in qualitative research projects. I have always seen journal writing as a major source of data. It is a data set that contains the researcher's reflection on the role of the researcher, for example. It is a great vehicle for coming to terms with exactly what one is doing as the qualitative researcher. Often, qualitative researchers are criticized for not being precise about what they do. I offer journal writing as one technique to accomplish the description and explanation of the researcher's role in the project. Qualitative researchers may use a reflective journal to write about problems that come up on a regular basis. Examples of problems include representation of interviews and field notes, coconstruction of meaning with participants in the project who also keep journals, and issues related to the interpretation of each other's data. Often, we qualitative researchers are positioned outside the very people and situations about which we write. Journal writing personalizes representation in a way that forces the researcher to confront issues of how a story from a person's life becomes a public text, which in turn tells a story. Furthermore, how are we to make sense of this story?

Basically, the art of journal writing and subsequent interpretations of journal writing produce meaning and understanding that are shaped by genre, the narrative form used, and personal cultural and paradigmatic conventions of the writer, who is also the researcher and participant. As Progoff (1992), my favorite teacher about journal writing, notes, journal writing is ultimately a way of getting feedback from ourselves. It enables us to experience in a full and open-ended way the movement of our lives as a whole and the meaning that follows from reflecting on that life.

Issues to be considered by the qualitative researcher include movement from the field to the text to the final, public research report and problems of interpretation, meaning, and representation. Interactive journal writing between researcher and participants is another way of understanding a given study. I have written earlier about journal writing (1999) and then wrote of the lengthy history of journal writing. All periods of history have benefited from journal writing. After all, journals are texts that record dreams, hopes, visions, fantasies, feelings, and innermost thoughts. Today, one can look to the incredible work of Edward Robb Ellis (1995). Ellis was a diary writer who kept a journal from 1927 to 1995. He was born in 1911 and kept a journal for 67 years, or more than 24,000 days. His descriptions provide amazing coverage of the events he lived through personally, but he also describes societal changes.

Yet, literary and historical figures are not the only journal writers. The field of psychology has long made use of journal writing as a therapeutic aid. The cathartic function of journal writing has been widely recommended by many schools of therapy. Therapists view the journal as an attempt to bring order to one's experience and a sense of coherence to one's life. Behaviorists, cognitive therapists, and Jungian analysts have used journals in the process of therapy. The journal is seen as a natural outgrowth of the clinical situation in which the client speaks to the self. Most recently, Progoff (1992) has written of an intensive journal. Progoff developed a set of techniques that provides a structure for keeping a journal and a springboard for development.

As a therapist himself, he has conducted workshops and trained a network of individuals to do workshops on keeping an intensive journal for unlocking one's creativity and coming to terms with one's self. The intensive journal method is a reflective, in-depth process of writing, speaking what is written, and in some cases, sharing what is written with others. Feedback is an operative principle for the Progoff method. The individual needs to draw upon inner resources to arrive at the understanding of the whole person. The journal is a tool to reopen the possibilities of learning and living. Progoff advocates the following:

1. Make regular entries in the journal in the form of dialogue with one's self.

2. Maintain the journal as an intensive psychological workbook in order to record all encounters of one's existence.

3. Attempt some type of sharing of this growth through journal writing with others.

The method makes use of a special, bound notebook or computer file divided into definite categories that include the following: dreams, stepping stones, dialogues with people, events, work, and the body. The writer is asked to reflect, free-associate, meditate, and imagine what relates to immediate experience. Currently, one only has to walk through the display aisles of the major bookstores, such as Borders or Barnes & Noble, to see the many examples of recently published journals. Recently, I found the following:

- The journal of an Iraqi war survivor
- The Andy Warhol journals
- The journal of a surgeon
- The journal of an army veteran

The point is that this genre is alive and well, and qualitative researchers should not be afraid of trying to keep journals. In fact, journal writing is so prevalent now that one only has to surf the Internet to see thousands of journal resources, examples, and personal histories online. For example, there is an online course on journal writing offered by Via Creativa, a website entirely devoted to Progoff's Intensive Journal Workshop, chat rooms on journal writing, exemplars of diaries and journal writing, and literally thousands of resources. In general, the common thread that unites all of these resources on the Internet is the agreement that journal writing is a way of getting in touch with yourself in terms of reflection, catharsis, remembrance, creation, exploration, problem solving, problem posing, and personal growth. I see all of these as part of the research process. For qualitative researchers, journal writing offers a way to document the researcher's role, triangulate data by entering the journal itself as a data set, and use the journal with participants in the study as a communicative act.

Why Journal Writing?

Students and colleagues have often asked me why they should invest time in journal writing. To this, I can only reply that journal writing allows one to reflect—to dig deeper, if you will—into the heart of the words, beliefs, and behaviors we describe. It allows one to reflect on the tapes and interview transcripts from our research endeavors. If participants also keep journals, it offers a way to triangulate data and pursue interpretations in a dialogical manner. It is a type of member check of one's own thinking done on paper. The clarity of writing down one's thoughts will allow for stepping into one's inner mind and reaching further into interpretations of the behaviors, beliefs, and words we write. Not everyone finds it easy to keep up with the demands of journal writing. The discipline and desire involved nearly outweigh some individuals' abilities or time. On the other hand, can this not be an option for all who are interested in becoming better researchers, writers, thinkers, and scholars? How does one set time apart for journal writing? I recall the teacher who said she had only 20 minutes after school to write in her journal, and that was that. Then, she ultimately decided she needed to keep a journal at home as well, because once she started to write, she found that she was staying at school and writing for at least an hour each day. She got up an hour earlier than anyone in her house and started writing in the early morning hours, a technique advocated by many writers. Many writers of journals have directly or indirectly stated how journal writing can assist one in developing creativity. The focus and energy demanded of one who writes a journal can be instructive for qualitative researchers. See Ranier (1978) for her categories below:

1. *Travelers.* People who keep a written record during a special time, such as a vacation or a trip

2. *Pilgrims.* People who want to discover who they really are

3. *Creators.* People who write to sketch out ideas and inventions in art or science

4. *Apologists.* People who write to justify something they have done and plead their case before all who read the journal

5. *Confessors.* People who conduct ritual unburdenings with the promise of secrecy or anonymity

6. *Prisoners.* People who must live their lives in prisons or who may be invalids, and as a result, must live their lives through keeping a journal

Of course, any writer might be a combination of any of these categories, but this might be useful as a tool to understand different approaches to keeping a journal. Progoff (1992) gives numerous examples of individuals who fall into these categories to illustrate the importance of keeping a journal. In fact, he got interested in writing his book because he himself has kept a journal for more than 30 years. I share that interest with him.

Rainer's (1978) book, *The New Diary*, contains superb examples of journal writing. I agree with her use of the terms *journal* and *diary* interchangeably. She describes seven techniques for journal writing, some very similar to Progoff's techniques. Her list is one that qualitative researchers may recognize as being used regularly in the arts and humanities.

Rainer's Seven Techniques

1. *Lists.* This technique allows for a person to write lists of activities, such as things to do, things that upset a person, things that are problematic, and so forth. It allows a writer to capture the pace of his or her activities and can be a good beginning for a journal writer who may go back and fill in the story in narrative form regarding all of the entries on the list.

2. *Portraits.* This allows the writer to describe a person or any number of people. The portrait is never really finished for the qualitative researcher. It evolves and takes on a life of its own throughout the project, and the writer may add to and subtract from it as the work takes shape.

3. *Maps of consciousness.* This technique is borrowed from the arts, and it involves actually drawing a map of what one is thinking. Rainer advocates using stick figures, lines, or shapeless blobs. It is a way to free up one's thoughts and put them to paper in another format.

4. *Guided imagery.* This technique is borrowed from the field of psychology, which advises that daydreaming images allow for an individual to start writing about any given topic.

5. *Altered point of view.* In this technique, the writer takes a different perspective on any given activity. For the qualitative researcher, for example, one might write about something in an observation or interview from another person's viewpoint, not the researcher's viewpoint. Many beginning researchers find it hard to write in the first person, and they talk about their projects in a third-person voice. It is a way of looking at something from the outside. For Rainer, looking from the outside might aid in getting to the inside of a topic.

6. *Unsent letters.* Obviously, this is about writing a letter to someone without any intention of showing it to that person. In a research situation, the researcher may write to one of the participants in the study, for example.

7. *Dialogues.* This is the technique Progoff suggests, and many writers use this effectively.

As I finish this section, the major ideas I want to punctuate have to do with journal writing as a technique used in the arts that resonates with the qualitative researcher. Writing down what we think and feel helps in the journey to improve our research practice, for example. Some of the personal examples used in the body of this text may serve to illustrate the individual writer's thinking processes and the willingness to analyze, rethink, and go deeper into a critical stance about one's life and work. Progoff calls this the scope of personal renewal. Others call it reflection. Still others, myself included, see journal writing as a tangible way to evaluate our experiences, improve and clarify our thinking, and become better writers and scholars. The researcher reflective journal, used while conducting qualitative research projects, helps the researcher with the following:

1. Helps to focus the study

2. Helps set the groundwork for analysis and interpretation

3. Serves as a tool for revisiting notes and transcripts

4. Serves as a tool to awaken the imagination

5. Helps to keep the written record of thoughts, feelings, and facts

We are talking about examining our own thoughts, beliefs, and behaviors. Many will say that helps only the writer. Still, if that were the only outcome of writing a journal, I would say that this continuing self-reflection is a first, vital step in modeling this for our students. Journal writing is a powerful research technique for the researcher and the participants in a given study. The definitions of the roles of the researcher and participants in a study are clarified through the reflection and the writing process involved in journal writing. Because the researcher is the research instrument, keeping a journal is a check and balance during the entire course of a qualitative research project. Likewise, keeping a journal during the course of a research project is a way to practice interdisciplinary triangulation. In order to get students to begin this project, I suggest the following exercise.

An Activity to Enrich Your Researcher Reflective Journal

Think about the points listed here before you begin this activity.

- Examining the familiar
- Observing the details
- Exploring positions
- Investigating issues
- Revealing current thoughts
- Reflecting your own direction

1. Write one sentence that indicates something you already know about journal writing.

2. Write one sentence that indicates what you would like to know about journal writing.

This gets things started.

Next, we have volunteers read to the class their first sentence and then their second sentence. Then, we begin a journal writing exercise.

Exercise 5.6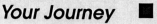

Your Journey

Position yourself in this present moment. Describe your journey to this class. How did you come to this class today? This program? What prompted you to select this area of study? What were your obstacles? Expectations? Write about yourself and your goals. How can this class make you a better researcher?

As you can see, this inspires quite a bit of writing. Next, we pair up and read to one another for 30 minutes. When we return to the larger group, volunteers read sections of their first journal entries. Learners are delighted that many in the group have similar interests and backgrounds. Many are inspired to write more. But overall, everyone writes and rewrites. In the next class meeting, we continue to save 15 minutes per class for journal writing.

At the end of each session, I ask all to do the following:

Please write three things that you learned about journal writing today.

This keeps learners focused on writing, writing, writing. In fact, students tell me that they run home and continue to write in their journals. For many, it is a liberating activity. (See Appendix B for multiple examples of journal writing.)

■ Future Directions

In speculating on the future of this useful technique of journal writing, I think that researchers in training may benefit from the practice of journal writing as a qualitative research technique for the following reasons:

1. Journal writing allows the writer to be more reflective.

2. Journal writing offers the writer an opportunity to write without interruption and to be totally focused on the point at hand.

3. Journal writing is a technique well used in the arts and humanities that may offer social science researchers an opportunity to cross borders, so to speak.

4. Journal writing allows for deepening knowledge.

5. Journal writing allows participants in a research project an active voice.

6. Journal writing may allow researchers and participants the opportunity to write cooperatively and interactively as needed.

7. Journal writing provides an additional data set to outline, describe, and explain the exact role of the researcher in any given project.

In concluding this section, I recall these words for thought:

There was so much to write. He had seen the world change; not just the events; although he had seen many of them and watched the people, but he had seen the subtler change and he could remember how the people were at different times. He had been in it and he had watched it and it was his duty to write of it.

—*Ernest Hemingway*

The Snows of Kilimanjaro *(1995)*

Exercise 5.7   ■

Reflective Journal Writing Practice
in Dialogue Form   ■

Purpose: To keep a journal in dialogue form over the length of
                the semester. Set aside a minimum of 15 minutes each
                day for writing.

Problem: The individual records minidialogues with the self
                in the present, focusing on the following
                areas.

People

Focus on dialogues with key people in your own life: best
friends, lovers, partners. What have you discovered
about the person you are today as a result of these
dialogues?

Work or Projects

Focus on a dialogue about significant projects or work that take up a
great deal of your energy. Which projects succeeded? Which failed?
What have you discovered about yourself as a result of these
projects?

(Continued)

(Continued)

The Body

Focus on your view of your own body. How have you cared for it? How do you treat it today? Are there moments in your life history where you mistreated your body? What have you discovered about yourself as a result of awareness of your body?

Society

Focus on your relationship to social groups. Of which groups are you a member? How do you describe your own ethnic and racial identity? Are you aware of your political beliefs? Are you reconsidering how you relate to groups? What have you discovered about yourself as a result of awareness of your relationship to society?

Major Life Event(s)

Focus on one or more life events that have had a profound effect on the person you are today. What have you discovered from reflecting on this?

For many learners, keeping a journal is a new experience, and by reading about journal writing and doing it as an activity, learners begin to reflect more deeply on their roles as researchers and as human beings. The journal is kept throughout the class to show evidence of one's progress on the journey of developing as a qualitative researcher. It is an opportunity to think in a new way. In addition to keeping a journal, the student is asked to include some poetry indicative of or using one of the techniques practiced in class.

Exercise 5.8 ■

Haiku and Any Form of Poetry
on the Role of the Researcher ■

Purpose: To write a poem in any style, including haiku, about the
role of the researcher or one that reflects some skill in
describing an object, place, person, or setting.

Problem: To capture the essence of the individual's role in a
particular study undertaken during the semester or as
part of a larger project.

Activity: The student is introduced to haiku, which is 17-syllable
Japanese poetry in its classic form with 5–7–5, or 5
syllables in Line 1, 7 in Line 2, and 5 in Line 3. Also
introduced is the 14- to 17-syllable form. Haiku is the
poetic form most like qualitative work because it takes
its imagery from careful observations. Many complain
that they are unable to write poetry, so I give them
some of my own samples of haiku as well as other
students' work. Amazingly, they seem to feel
exhilarated after seeing that they are, indeed, able to
do this. Recently, students took up writing poetry on
their own about doing their dissertation projects. So, I
expanded this exercise to include any form of poetry.
The point is to capture in another idiom—poetry—
something of what occurred in the study, either about
the role of the researcher or the participants in the
study. Here are some examples.

(Continued)

(Continued)

Poems Written While Experiencing the Writing of the Dissertation

By Ruth Slotnick (2010), doctoral candidate in higher education

Uncovering

Draped in nuances
where lines blur
suddenly become clear
revealed in utterances
where truth arises
the themes emerge
and transformation ensues
A balance is restored
life deepens; enriches as
possibilities unfold
This is a process
no number can reveal
what shadows conceal—
stirred out of dormancy
waiting to be told
This is our passion
Our destiny

When Analysis Eludes (on Writing the Dissertation)

Foggy place
Burry space
Clarity desired
Tired and mired

Not a number
Nuances asunder

Unable to unscramble
Barbs and bramble

Unpuzzled daily
Though today, not gaily
Start over, retract
Intuition out of whack

Emerge, emerge!!
Frustration, purge
Constructing lines
Weave symbol and sign

Analysis Breakthrough

Rummaging
Repeatedly
Revisiting
Rhythms
Revising
Realities
Reasoning
Representations
Reframing
Results

By Pat Williams-Boyd, Professor, Eastern Michigan University

This Woman

A woman of many seasons sits on a corroded metal lawn-chair
as a sentinel protecting her wealth.

Her head is covered with a frayed babushka, tied securely under
her chins, refuting the

(Continued)

(Continued)

windless 90-degree California sun.

Her unmatched men's shoes are tied on with black laces, and

her stockings are rolled down to her ankles.

Her simple blue and white cotton wash dress sparkles in the

reflection of the piercing brown eyes, which

study everything yet nothing from behind a weather-beaten face.

Her hands, gnarled yet powerful, rest calmly in her bountiful lap.

A tabby cat sits on her shoulder as if an epaulet

adorning her brown, mended, long-sleeved sweater.

I have watched her unnoticed from a distance for some time.

She has neither moved nor altered her gaze

from the red brick house, which sits brazenly across the street.

Where her yard is blanketed by sand and debris,

her neighbor's is lushly green, splashed with

rosebushes and October pumpkin designs.

But, I sense her vision is not of this world

separated by so random a stretch of asphalt

but of a world in which she was part of the earth.

Her eyes are of a distant time, and she sees neither me, nor her dreams.

I am not here.

Exercise 5.9 ∎

Framed Photograph and Narrative Writing Exercise ∎

Purpose: These two exercises combine description and narrative writing. Describe a framed photograph of a familiar person. Then, pair up with someone and exchange photographs, and each of you describes the other person's selected framed photograph. Then, write about your partner's photo as though you were a reporter telling that person's story.

Problem:

1. Individuals think about and use time to describe the framed photograph they themselves bring to class.
2. Next, they pair off and exchange their own framed photographs with their partners. After receiving the partner's photograph, each describes the partner's photograph.
3. Partners stop to share and discuss their descriptions. Each group of two has two descriptions to review and find differences and similarities. This is designed to prepare learners for analysis of data exercises later.
4. Next, everyone stops to write about this as if they were reporters trying to tell a person's story. This writing entry is of course to be considered for the researcher reflective journal.

Time: At least 15 minutes for each of the two descriptions. At least an additional 15 minutes for discussion. At least 30 minutes for writing and rewriting.

Activity: This activity serves to sharpen awareness of the role of the researcher by working with a familiar artifact, a

(Continued)

(Continued)

framed photograph of someone dear to the learner. At the same time, it extends and reviews observation expertise by moving to the description of an unfamiliar photograph. By pairing off, learners have the opportunity to practice communicating with another researcher about something in common, which is how to approach both descriptions and how to compare and contrast descriptions, thus establishing a collaborative habit. This is a precursor to the next round of exercises, which include practice in interviewing individuals and analyzing interview data. In addition, the writing exercise continues the habit of writing and, beyond that, writing for the researcher reflective journal.

Although, strictly speaking, this exercise forces the individual to observe and could fit in the observation cycle, this activity allows for a transition to interviewing and analysis of data and focuses on discovery of the role of the researcher and writing in the reflective journal.

Discussion: Following the exercise, individuals are asked as a group to respond to the following questions:

1. How did you approach the first part of the activity? The second part?
2. What differences did you find in your own thinking as you approached each description?
3. What was the most challenging part of the exercise for you?
4. Would you read a sample of your description?
5. Could you write in your researcher reflective journals on this experience?

Rationale: This exercise again is part of the practice of disciplined inquiry designed to make the researcher more reflective, establish and strengthen writing habits, and encourage collaboration. Reflection on the actual

(Continued)

mechanics of approaching the description of the photographs is the first part of the activity. In the second part, individuals are forced to reflect on their own roles as researchers. By discussing this with other people, members see similarities and differences as part of the scope of disciplined inquiry. The habit of writing in the researcher reflective journal and contributing to the portfolio is enhanced. Creativity is the underlying theme in all these exercises. Creativity is a habit that is practiced and learned.

■ Summary of Chapter 5: The Creative Habit

Due to the fact that the researcher is the research instrument in a qualitative research project and that creativity and intuition are part of the role of the qualitative researcher, these exercises have been designed to sharpen one's creative habit and awareness of the role of the researcher. This can be traced to the 17th-century text, *The Mustard Seed Garden Manual of Painting*, which was part of a larger work, *The Tao of Painting*, by Lu Ch'ai. Nearly everything written by the Chinese master painters was aimed not just at the technique of painting but also at the painter's spiritual resources in order to express the spirit, or *chi*, the breath of Tao. The *chi* is looked upon as an underlying harmony. Likewise, in dance, the spirit of the dance must emerge as part of movement. These exercises are akin to the *chi* meaning of painting and the spirit of the dancer's movement. In order to stretch, the qualitative researcher must be able to articulate creativity and the role of the researcher as the underlying harmony or spirit of the study. The qualitative researcher is always dealing with lived experience and must be awake to and for that experience. By acknowledging and articulating the complexity of need for creativity, defining the role of the researcher, and keeping a reflective journal, we are now able to begin the next cycle of exercises in the next series, the analysis cycle.

■ Resources and Readings

Berg, B. L. (1995). *Qualitative research methods for the social sciences* (2nd ed.). Boston: Allyn & Bacon.

Bing, S. (2002). *Throwing the elephant: Zen and the art of managing up.* New York: Harper.

Dali, S. (1986). *The secret life of Salvador Dali.* Wantagh, New York: Geneva Graphics for Dasa Edicions.

Edwards, B. (1979). *Drawing on the right side of the brain.* Los Angeles: J. P. Tarcher.

Gelb, M. J. (1998). *How to think like Leonardo da Vinci.* New York: Dell.

Hemingway, E. (1995). *The snows of Kilimanjaro.* New York: Scribner.

Janesick, V. J. (2001). Intuition and creativity: A pas de deux for the qualitative researcher. *Qualitative Inquiry, 7*(5), 531–540.

Kohler Reissman, C. (2008). *Narrative methods for the human sciences.* Thousand Oaks, CA: Sage.

Leavy, P. (2009). *Method meets art.* New York: Guilford Press.

Mallon, T. (1995). *A book of one's own: People and their diaries.* St. Paul, MN: Hungry Mind Press.

Nin, A. (1976). *The diary of Anais Nin, 1955–1966* (G. Stuhlman, Ed.). New York: Brace.

Noddings, N., & Shore, P. J. (1984). *Awakening the inner eye: Intuition in education.* New York: Teachers College Press.

Progoff, I. (1992). *At a journal workshop: Writing to access the power of the unconscious and evoke creative ability.* Los Angeles: J. P. Tarcher.

Rainer, T. (1978). *The new diary.* New York: G. P. Putnam.

Richardson, L. (1995). Writing-stories: Coauthoring "The Sea Monster," a writing story. *Qualitative Inquiry, 1*(2), 189–220.

Tharp, T. (2003). *The creative habit: Learn it and use it for life.* New York: Simon & Schuster.

Von Oech, R. (1986). *A kick in the seat of the pants.* New York: Harper & Row.

CHAPTER 6

The Analysis and Writing Habit ■

Making Sense of the Data, Ethics, and Other Issues ■

Research is to see what everybody else has seen and to think what nobody else has thought.

—Albert Szent-Gyorgi (1996), Biochemist (1893-1986)
1937 winner of the Nobel Prize for Medicine

Returning to the dance metaphor, if these exercises represent the stretching in a dance class, analysis of data and interpretation and representation of data represent the floor exercises and performance stages of dance. After completing the series of exercises in observation, interviews, and the role of the researcher, learners have stretched a bit. They've had a taste of some aspects of data analysis through study of interview transcripts and field notes. Working either alone or in groups, learners have practiced developing categories from the data, looking for points of tension and conflict, and in general, focusing on making sense of the data from the entire term. As Wolcott (1995) pointed out, there is a difference between analysis and interpretation. In order for prospective researchers to realize their interpretation skills, it is important to think about the use of intuition in research just as it is used in choreography.

The role of the qualitative researcher in research projects is often determined by the researcher's stance and intent, much like a historian. Likewise, choreographers create dances with the knowledge of where they fit in the history of dance. In addition, the roles of the researcher, like the dancer and choreographer, are ones where all senses are used to understand the context of the phenomenon under study, the people who are participants in the study and their beliefs and behaviors, and of course the researcher's own orientation and purposes. Furthermore, it is like the ending in an O. Henry story. In other words, it is complicated, filled with surprises, and open to serendipity, and it often leads to something unanticipated in the original design of the research project. At the same time, the researcher works within the frame of a disciplined plan of inquiry, adheres to the high standards of qualitative inquiry, and looks for ways to complement and extend the description and explanation of the project through multiple methods of research, providing that this is done for a specific reason and makes sense. Qualitative researchers do not accept the misconception that more methods mean a better or richer analysis. Rather, the rationale for using selected methods is what counts. The qualitative researcher wants to tell a story in the best possible configuration.

■ The Qualitative Researcher as Historian

The qualitative researcher is most like a historian in that special access to sources is critical. When appropriate in research projects, the qualitative researcher relies on many possible sources of data and uses a variety of methods in the process, including but not limited to observation, participant observation, interviews, documents, the researcher's personal reflections, and so on. From the conceptualization of the research project to its completion, the researcher needs to be direct in terms of identifying bias, ideology, stance, and intent. It is usually the case that the qualitative researcher wants to understand the situation under study and must decide if the stance is taken from the inside or outside, as participant or observer, or as some combination with varying degrees of both, and whether or not the researcher will approach the project holistically. Like the historian, the qualitative researcher must choose a point of view, either as an apologist for what occurs or as the intimate insider, looking at the whole and explaining each part of the puzzle. These are the crucial questions that need to be dealt with before entering the field. It is wise to pilot test observations, interviews, and any type of participant observation that is being considered for the research plan. By doing a pilot study of limited duration, one is able to get a taste for the setting, become acquainted with personnel, and test one's skills as observer and interviewer. Just like a historian, then, the qualitative researcher uses primary sources either as an insider and participant or as an outsider. As participant, one has the advantage in terms of special access to sources. As a role model, we might look to Thucydides, for example, who did his own writing and data collecting. More recently, we might look to the great historian Allan Nevins, who pioneered oral history methods and constantly demonstrated the importance of the insider's view of the social setting and its participants. In any event, each of us has to declare and describe our points of view, our theoretical frames, and our beliefs about the topic we are researching. Most often, this is done in the description of the role of the researcher.

■ Checkpoints for Data Analysis, Reporting, and Interpretation

1. Look for empirical assertions supported by the data. Look at what the participants said in the transcripts to find those meaning units.

2. Use narrative vignettes and exact quotations from participants to support your assertions. In this case, more is more. Convince your reader of your argument with evidence from transcripts, observations, reflective journals, and any other documentary evidence.

3. Scan all other reports, documents, letters, journal entries, demographic data, and so on. Use direct references. Connect these to your analysis and interpretation.

4. Include interpretive commentary related to the data, because data simply do not speak for themselves. In other words, lead your reader to your themes.

5. Include a theoretical discussion, and relate your data to the theory that guided the study. Give your readers a hint of what will be included in your model of what occurred in the study.

6. Be sure to include a thorough description of your role as the researcher as part of the analysis and history of the inquiry, and include your beliefs on the topic.

7. State clearly any and all ethical issues that arose in the study. What these checkpoints do is allow your reader to experience the study and its social context and setting.

8. Have a peer reviewer, outside reader, or auditor review your transcripts and look at your preliminary and final categories. To verify this, design a simple form for your peer reviewer stating that the person reviewed transcripts, notes, and the final write-up of the research and that there is a fit with the data.

9. Work for saturation and sustainability of your data. This means that you get to your goal in the project by saturating in the interview data, the relevant documents, your research reflective journal, and any other artifacts in the study, like photographs, for example. You will know the moment you get to your goal. There is no formula for that.

The range of evidence should be used to support your assertions and view the work in progress throughout the entire history of the inquiry. See Appendix M for a more thorough example of the process Carol Burg (2010) used in her dissertation on mentors' views of mentoring. She took the transcript itself, looked for meaning units, and then put those meaning units, or codes, into her software, Atlas Ti. She arranged four columns:

1. The transcript,

2. The theme of the meaning unit,

3. The emergent central theme, and

4. The emergent code.

Viewing this from a phenomenological point of view, she then took the emergent code and put that into Atlas Ti to see if anything else would come of it.

■ Example of Using a Transcript to Get to a Code

1. *Transcript:* Professor Intentional states: "I came across the term in my reading, intentional mentoring . . . that really kind of embraces my philosophy of mentoring. . . . It is intentional; it is conscious. Now, there are many things that happen in a mentoring relationship that

you don't think about consciously. It is serendipity. It is something that happens in a hallway. It happens in passing conversation."

2. *Theme of meaning unit:* Professor Intentional sees her role as intentional.

3. *Emergent central theme:* Mentors may see their role as intentional.

4. *The emergent code:* Mentoring: Intentional.

Now, look at another way to approach the analytical process. Ruth Slotnick (2010), as a social constructivist, decided to write about her analytical process in her researcher reflective journal. She uses a narrative approach and eventually constructs a model of her three cases (see Figure 6.1).

Figure 6.1 The Three Cases

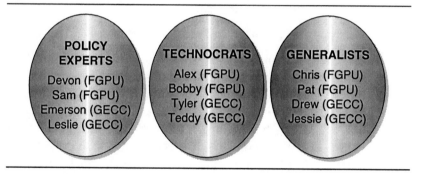

POLICY EXPERTS
Devon (FGPU)
Sam (FGPU)
Emerson (GECC)
Leslie (GECC)

TECHNOCRATS
Alex (FGPU)
Bobby (FGPU)
Tyler (GECC)
Teddy (GECC)

GENERALISTS
Chris (FGPU)
Pat (FGPU)
Drew (GECC)
Jessie (GECC)

■ Example of Thinking Through an Analysis

July 26, 2009: Analysis (Ruth Slotnick)

How much of the data corpus should be coded?

Some believe all data should be coded to locate the nuanced patterns; others believe that only the salient points should be coded, that most of the data can be excluded, leaving the quarter of the data to be analyzed. Saldana (2009) calls this "the primary half for intensive data analysis" (p. 15). He concludes by saying that, no matter which side you ascribe to, make sure your data is quality. Code trivial data as nonapplication or N/A (p. 15). He talks about "metadata activities" as codes that come about from field notes, observations, and analytical memos. Also,

Saldana recommends graphic summaries (p. 16). For analysis, he suggests dividing the text up into units according to the changing of topics. This will make analysis easier in Atlas-Ti. I will use topic and subtopic breaks as recommended. I will reformat all the transcribed texts to read this way. Saldana calls these unit divisions.

So my system of analysis is this:

1. Listen and make corrections to transcriptions. Although my transcriptionist has done a fairly good job of catching the text as spoken, she does not transcribe some of the nuanced expressions used, so I have filled in what was missed, for example, the sound of laughter, when appropriate. The transcripts with my corrections and additions are saved in an interview file.

2. Send file to interviewee for member check.

3. Make any changes to the document postmember check.

4. Delete any extraneous talk (regarding the digital recorder and final good-byes) as well as other irrelevant text, such as extra ums or repeated words as a sentence is being formed; "I . . . I thought I might . . . I thought I might be doing the wrong thing" would look like "I think I might be doing the wrong thing." The digital recording will help me determine if there is hesitation in the voice of the interviewee. This will be noted in parenthesis. This becomes the draft from which I will code.

5. The next step is to start summarizing the interview text as recommended by Rubin and Rubin (2005), summarizing each passage with a few words and looking for themes, concepts, events, and topical markers.

6. While the document is still single-spaced, break texts into topics and subtopics. Saldana (2009) calls them stanzas that help to format the text and reveal meaning and intent otherwise lost in long paragraphs of text (p. 16). Breaking the text into stanzas is also key to make coding in qualitative software easier to compile and analyze.

7. Use a left-flush margin, leaving a 2-inch, right-hand margin for pencil coding.

8. Precoding: With digital recording, field notes, and the researcher's reflective journal, I will begin my precoding. Precoding is where

I am circling, underlining, bolding text. These are also called "codable moments" (see Saldana, p. 16).

9. Preliminary jotting: The goal is to have all the tangible pieces related to each case available for consideration as the analysis commences.

10. Primary coding: This is the first round of identifying codes.

11. Secondary coding: In this stage, I'm building categories.

12. Third round: Then I finalize categories and cases.

September 6, 2009: Analysis

What am I hearing? What am I not hearing?

What assumptions am I making or are my interviewees making?

What is being downplayed or overplayed?

What policies and practices enhance or increase the chance of access and student success?

What are external pressures? Internal conflicts?

Developed a case study matrix. Further developed into three cases.

September 9, 2009: Analysis

I am developing a proximity matrix. Let's see what differences I find! I think these will either be responses that are more subtle or that indicate a general lack of familiarity or brevity with the state policy. I created instead a policy proximity continuum using *policy expert* as the closest description and *technocrat* as slightly further away.

September 10, 2009: Analysis

I created continuum for all three cases. Then I met with Valerie, my methodologist. She was happy with my progress. We came up with suggestions for the subcategories for close and distant on the proximity continuum. I went home and reworked this continuum. Then, I started to work on the next category: perceptions of underrepresented students, differentiated and undifferentiated.

End of journal entry

Thus, it is evident that there is no one way to do data analysis. However, the tried-and-true method of anthropologists, sociologists, historians, and critics is helpful here:

1. Pour through your written and audio data.

2. Select themes and codes based on frequency, distribution, and meaning to the participant.

3. Read the transcripts over and over again. After you get a transcript, it is good to read it a few times or more per week as you move into the analysis and interpretation phase.

4. Write everything down that you are thinking of, and write about what this means.

The data tell you something. Your job is to find out what the data are saying to you. Now, you constantly compare and contrast what you are finding and come up with some interpretation of all that you compared: themes, meaning units, words, and categories about the themes. Also, as you look for repetition and constant recurring statements, you know you have a grip on what the participant is telling you. Next, you need to interpret these data. Make a visual diagram or figure representing your data and what you found. In addition, the range of ethical issues encountered in a study should be part of its analysis and interpretation. Qualitative researchers should include a section on ethical issues that arose in the study in any of their reports and especially in the dissertation.

■ Ethics and the Qualitative Researcher

If the qualitative researcher can be assured of serendipity and contradictions, he or she can also be assured of ethical issues. In conversations about works in progress, dissertations, and class projects, learners need a forum for discussing the ethical issues that arise in any given field setting. To prepare for discussing these everyday moments of fieldwork, learners have an opportunity to discuss and react to actual cases of ethical dilemmas that previous students have encountered. To warm up the group in discussing ethical issues, learners assemble in groups and choose one of the following actual cases for discussion. Think of these two exercises as stretching your thinking about ethics and fieldwork.

■ True Stories: Sample Ethical Dilemmas

Discussion Exercises

The Case of "Just Delete That Data"

Recently, a doctoral student was conducting interviews on tape for a study that was designed to describe and explain the quality of a federally funded music and arts camp for the handicapped, with a particular focus on teacher effectiveness. The seven teachers were teaching music, art, piano, voice, and composition. The student conducted interviews with all of the teachers over a 6-month period and observed in the seven classrooms. He also transcribed all the tapes and found that one teacher was totally neglecting his work as an instructor. Not only that, all of the other instructors knew that one of their group was doing this and that he was basically unqualified, rude to students, and arbitrary in his assessments. The researcher showed these data to the director at a member check discussion; the director was also one of the teachers. In the discussion, the director revealed that he hired that teacher and could not accept the results of student evaluations or the comments from other instructors. The teacher in question was an old friend. Furthermore, the director asked the researcher to delete all comments referring to that teacher in case the data jeopardized funding for the next summer.

Questions for Discussion of the Case

1. How would you handle this?

2. What is the role of the researcher in this case?

3. What ethical issues are raised in this case?

4. What would you change, if anything, in this situation?

The Case of "I Never Said That"

A doctoral student was interviewing female administrators over a 2-year period about their perspectives on their roles as women in administration. All interviews were taped. In the course of one interview, one of the high-level administrators disclosed that, as a child, she was abused by her father, who was an alcoholic, and that she had decided not to have children lest she inflict anything on them that she had once experienced. Later, when the researcher met with this administrator to conduct a member check and discuss the write-up of the case study, the administrator said, "I never said that." The researcher tactfully pointed out that she had said that and that it was on tape. Once again, the administrator denied this and threatened to drop out of the study if the researcher even hinted at this in the case study write-up.

Questions for Discussion of the Case

1. What ethical issues are raised by this case?

2. What would you change, if anything, in this situation?

"I Said That, But Now I Want It Erased"

As you know, anything can happen in the social world. That is why we enjoy this type of research. Recently, a doctoral student got incredibly rich data on tape from a nurse educator who was onto something like the Enron fiasco. The nurse educator read the transcripts, signed that she read them, and after the study was completed, returned to the researcher and said that she would like to delete some of the things she said because she was fearful of repercussions from her supervisor. The researcher was now faced with a decision to (a) recast what was said in a way that would be less threatening, (b) delete the information entirely, or (c) leave everything as it was. What would you do in this instance?

Of course, there are endless examples of ethical questions in qualitative research studies; these three cases offer some examples for discussion. Actual examples each semester provide grist for the forum, and by using group members to share viewpoints, prospective researchers at least begin to grapple with ethical issues and write descriptions of these issues in their reports. Likewise, doctoral students need to write about and include ethical questions that arose in their studies as part of the ongoing explication of all components of qualitative work. In dissertations, writers often include the ethical questions in the Methodology section under Role of the Researcher or in an appendix, as needed, depending on the nature of the ethical issue.

■ Other Issues

Throughout my experience, whether at public presentations of this material or in the classroom setting, a number of questions regularly arise. As a way of responding to those questions, I have written earlier (Janesick, 1998) of rules of thumb that guide our work. As you look at these rules of thumb, you will easily identify the questions that resulted in this heuristic tool.

■ Rules of Thumb for Qualitative Researchers

Try to refrain from studying your own group. For many years, researchers have warned against this. In class, students who have studied their own groups have found the difficulties and anguish far greater than any benefits. As a member of a group, you may be too close to that group to be fair and accurate in your reporting. In addition, in hindsight, students who have studied their own schools or their friends' schools have sometimes been turned down for jobs and advancement in their current work situation due to someone's fear of research.

Always have an outside reader or peer reviewer. As you may know, Malinowski, Powdermaker, Mead, Bateson, and other classic anthropologists used outside readers of their field notes. Novelists and playwrights also use outside readers to bring a fresh viewpoint and to read for discrepancies and the like. Especially as a researcher in training, it is wise to use an outside reader of your field notes and interview transcripts. Other writers (Lincoln & Guba, 1985) suggest the use of an audit trail from the field of business. This is a fine idea if one has the time and extra money. However, as one who relies on the arts and humanities for my history and being, I prefer the use of the outside reader because it offers a long and dependable history in our field. Lately, students in my doctoral classes have gotten written verification from their outside readers and are calling them peer reviewers (see Appendix K). This helps the researcher all the way around. The notion of a peer reviewer is rooted in history and can only add to the final product, the report of the study. This is our history. Can we all agree to honor it?

Design your study to understand. Qualitative work demands that the researcher avoid trying to prove something. Instead, the heart of our work is understanding the social setting and all that it entails. This means that you do not go into the field with the answers. You are always framing questions. It also means that you do not go into the field empty-headed either. Before going into any setting, you should already know about yourself as an interviewer, an observer, and a writer, specifically from your pilot study. Often, neophyte researchers appropriate jargon from another paradigm and try to recast qualitative work in another ill-fitting image. Qualitative work just does not work that way. We do not go out to prove something or control something. We go to understand something. The more a prospective researcher can read and practice, the better off he or she will be. Build up your library on method and read and reflect upon these texts. Recently, I asked successful doctoral students—those that finished—what factors contributed to the completion of their projects. Every single person listed the fact that they went out and purchased every text that was of interest to their approach to research among other factors.

Time in the field equals time in analysis. I have always agreed with those who hold to this guideline. If you spend a year collecting data, expect to spend a year analyzing them. The large amount of text to pore through demands a thorough and just accounting of the data. Many students short-change their participants by trying to get done with the study as quickly as possible. This is a sad testimonial to our field if this is allowed to prevail. You cannot do justice to your long-term study or your participants if you run in and out to meet arbitrary personal deadlines or some in-house deadline for graduation. Unfortunately, students in education often are so desensitized to their own roles in their organizations that they prefer to let the organizations rule them rather than the other way around. In all my cases of working with doctoral students, I tell them up front that the path is long. They must do equal time in analysis and in the field. Otherwise, we cannot come to agreements. In addition, if a doctoral student wants to rush through and finish quickly for the wrong reasons, it shows. For example, a student recently wanted a particular job, and in her haste to finish, the final chapter looks like a rush job. Do you really want everyone reading *Dissertation Abstracts* to see a study that is less than complete? The rigor and high standards of qualitative research must be ever present.

Develop a model of what occurred in your study. In qualitative work, theory is grounded from the data: the words of your participants, your field notes, transcripts, reflective journal entries, and other written records. By developing a model of what occurred, the reader of the report is more able to make sense of the data and follow the researcher's argument. This also takes the report to another analytical level. Usually, for the dissertation, this is completed in the final chapter.

***Always allow participants access to your data through a member
check.*** As qualitative researchers, we have an obligation to our participants
to allow easy access to field notes, journals on the research project, interview transcripts, and initial and final categories of analysis. In fact, this
should be built into the informed consent document. In most cases, participants are not very interested in this up to the end of the study. It is always
a good idea to give a copy of the completed report or a final copy of the dissertation to one's participants. Some researchers like to show the transcripts to the participants. It is up to you how you do this member checking.
Likewise, some doctoral students also give a small gift to participants. For
example, one of my former students gave a book from the bestseller list to
participants with a thank-you note. I never discourage this because it is
obvious that the experience of working on the research project together is
totally transformative for both the researcher and participants. In addition,
we have currently taken to using written verification from participants
regarding access to the data, which avoids misunderstandings later.

Look for points of conflict, tension, and contradiction. Looking for
what does not make sense in a study, what does not quite fit, and what
exposes points of conflict often yields amazing information and insight. As
the researcher goes through mounds of data, points of conflict offer a good
grasp of events and are fruitful points of departure for analysis and interpretation. In dance, we call it *looking for the asymmetrical.* In yoga, we call
it *looking for the imbalance.* Often, new researchers get so caught up in their
studies that they forget to look for what is out of the ordinary and what
does not fit. Again, having an outside reader and constantly checking
your own thoughts through the reflective journal process are most helpful.
The rush to finish should not be so strong that you forget to look for the
asymmetrical.

***Estimate your costs and time and then add some additional costs and
time.*** Inevitably, new qualitative researchers are amazed at the cost of transcriptions, duplication of sections of the report or dissertation, and supplies
and equipment. Estimate about $2,500 to $3,000 for completing a long-term
qualitative study for the dissertation, excluding tuition. Currently, transcribers are charging about $100 to $120 for a 1-hour tape. One hour of
taped transcript is approximately 21 pages of single-spaced transcription.
Some researchers negotiate with a transcriber for the cost of the total package. Tape recorders, tapes, and video recorders can be purchased used or new

at reasonable prices, but for example, a used video recorder is about $200. Likewise, time is your most precious and valuable commodity. Whatever your target date for completion, add another 6 months to give yourself a reasonable window for reflection and rewriting. Remember that students who do their own transcripts save a great deal of money. Furthermore, copy editors currently charge around $500 for a 300-page dissertation.

Recent estimates from students who completed interview studies show the difference in cost.

a. For a student who did 20 transcriptions personally, the final cost of the dissertation came in at just over $2,000, excluding tuition for dissertation hours.

b. For a student who sent 16 transcripts to a professional transcriber, the final cost of the dissertation came in at around $4,300, excluding tuition.

These guidelines are not meant to be all-inclusive, but they do respond to some of the most often-asked questions about the nuts and bolts of doing qualitative research projects.

Always do pilot interviews and observations. I cannot say enough about the critical nature of conducting pilot interviews and observations and even double-checking one's reflective journal. The idea of pilot studies is as old as the hills. Researchers do pilot studies or at least components of piloting techniques to sharpen their skills at interviewing and observation. From the pilot period, one can learn how to recraft questions in an interview or be reminded of nuts-and-bolts issues. For example, recently, a student went to do an interview but forgot to check the tape recorder and was held up by running to the nearest store for batteries. The recorder was dead because the batteries were dead. But beyond the nuts and bolts, the pilot assists the researcher in training in terms of confidence, clarity of thinking, and the knowledge base of qualitative research.

Write every day. Enough cannot be said about daily writing. Write in your dialogue journal; write in your researcher reflective journal; write narrative descriptions; and practice, practice, practice. Then, go back and rewrite. My own writing schedule is that, no matter what, I write every day from 5:00 a.m. to 7:00 a.m. Then, when I am on a deadline, such as the one for this text, I average 4 to 5 hours per day. I write best in the morning and find that I get more done in 2 hours in the morning than in 3 hours later in the day. When a big project is in front of me, I also write on weekends for a few more hours.

■ Writing the Qualitative Dissertation

Getting Started

My work was made lighter as a teacher by the informative text written by Piantanida and Garman (1999), *The Qualitative Dissertation: A Guide for Students and Faculty*. Students take to this text immediately, and it is a comfort to them to know that what they are experiencing in this process is validated. I engage students by once again reiterating the sections of the first three chapters of the dissertation to get started. Yes, they have read about this in countless texts, but it does not hurt to repeat the contents of the dissertation proposal. We require the first three chapters for the proposal. So, here is a handy outline. Notice also on the web that there are a number of sites with dissertation coaches, writing coaches, information on how to do a proposal, how to do a literature review, and so on. To name just two with all the information on the process of the dissertation that are helpful for students, go to www.dissertationdoctor.com and www.academicladder.com. This latter site helps new professors understand the publish-or-perish system of academia. Furthermore, you can also find websites for digital transcriptions of your interviews. To name just two, you will find help and fair market prices at www.casting words.com and www.productiontranscripts.com. In addition, you may also find information on software for assistance in data management on the web. Finally, you may find free software on the web that helps in the transcription process, such as the site for Express Scribe. All these tools are for your private tool kit. But first, you need to get your brain thinking in terms of a proposal for the qualitative research project.

■ Suggested Sections for the Qualitative Research Proposal

Due to the fact that I come from a tradition where the first three chapters of the dissertation are written as the proposal, I offer it here as the model practice. The value in this approach is obvious. Students have a familiarity with their literature and methodology, and they can clearly articulate their purposes and the questions that guide the study. I advise all students to begin by journaling every day to get to the heart of their topics. Also helpful is spending time daily in the library scanning *Dissertation Abstracts* and relevant journals. Here is the suggested model:

Chapter 1: Introduction and Purpose

1. Introduction and background of the study.

2. Statement of purpose of the study: The purpose of this study is to describe and explain selected mentors' perspectives on mentoring.

3. Exploratory questions that guide the study:

    a. What elements constitute this perspective?

    b. What variables influence this perspective?

Note: The value of exploratory questions is clear. They provide the researcher with an open and wide opportunity for final analysis of the data and are to be included in the final chapter of results, analysis, and interpretation. The researcher responds to these questions, which are the framework for developing a model of what occurs in the study. All of them are provided in the final chapter.

4. Theoretical framework that guides the study: The theoretical framework that guides this study is phenomenology.

5. Scope of the study: Include timeline, number of members who will participate, how you are using purposeful sampling, and a rationale for selection of participants.

6. Outline of the remainder of the study.

7. Definitions: If you have particular terms that need definition, please add them to this chapter.

8. Summary of the chapter and statement of what is coming in the next chapter.

Chapter 2: Review of Related Literature

1. To organize this chapter, make a schema, if you like. There is an example of a schema in Table 6.1.

Table 6.1 Schema of a Section of a Literature Review by Ruth Slotnick

Transfer Literature Integrating Policy Implementation Theories				
Year	Author(s)	Policy Implementation Researchers Cited	Topic of Study	Focus on Underrepresented College Students?
1974	Knoell, D., & McIntyre, C.	Dror, E. Wildavsky, A.	Planning Community Colleges	Yes
1985	Richardson, R. C.	Wildavsky, A.	Baccalaureate for Urban Minorities	Yes
1989	Richardson, R. C.	Wildavsky, A.	Minority Achievement	Yes
2000	Schaffer, L. J.*	Bardach, E. Dye, T. Sabatier, P. Yanow, D.	Implementing Transfer Policies	None
2008	Gonzalez, J. M.*	Elmore, R.F. Hill, H.C. Lipsky, M. Mazmanmian, D. Pressman, J. Sabatier, P. Wildavsky, A. Yanow, D.	Academic Tracking Systems	None

*Unpublished dissertations.

2. Provide historical background, if applicable.

3. Include purposes of the review of related literature and how it relates to the purpose of your study and your exploratory questions.

4. Acquaint the reader with your knowledge of the critical studies already published that relate to your study. You, the researcher, decide how many of these studies relate and what is the significant literature in your content area. You are the expert here. If you need to write about methodology literature, include it here.

5. Find the contradictions in the literature. Look for the themes and issues that directly relate to your purpose, questions, and research methods.

6. Find precedents in the literature for your work, if possible.

7. If you are doing a historical study, it is possible that your literature review may be embedded in the purpose of the study, and your design of this chapter may vary.

8. Work for an integrative review of your selected literature.

9. Look for recent work in the past 5 to 7 years but with an eye to the historical precedents of that work.

10. Summarize your review, and connect this to the next chapter to introduce your methodology.

Chapter 3: Methods Used in this Study

1. Provide an overview.

2. Describe your methodology.

3. Precisely state the rationale for your choice of methods.

4. Precisely describe the number of interviews, observations, and supporting documents to be used, and state that you will tape-record and transcribe all interviews.

5. Integrate your knowledge of the methodology literature here.

6. Describe your role as the researcher. Later, after data collection and analysis, go back and rewrite this section to describe accurately and precisely what you did.

7. Describe and explain your pilot study.

8. Construct at least the first set of interview questions. In fact, you may wish to combine some possible questions that could arise in your second interview should you do a second interview. Here, the Institutional Review Board (IRB) has not approved selected studies which had the statement, "I will construct questions for the second interview after I complete the first interview." So, get to know your IRB members, and find out what other proposers used as interview protocols.

9. Provide the timeline for interviews, observations, and analysis of the data. For example:

   Data collection: 6 months

   Member checking: 1 month

   Data analysis: 6 months

   Final draft of the complete dissertation with references and appendices: 1 month

10. Provide the rationale for your choice of participants, and describe informed consent.

11. Describe your methodological assumptions, how you will ensure trustworthiness and credibility. Will you have an outside peer reviewer of your transcripts, categories, and narrative? (Yes.) Describe how you will do member checks.

12. Describe how you will collect and analyze data. Also, describe any and all equipment used for data collection (e.g., tape recorders, video recorders, photographic equipment, software, a particular model, etc.).

13. Describe how you will handle any ethical issues that arise in the study. Include a copy of your consent form. Describe how you will get IRB approval.

14. Add any appendices to your proposal as needed.

Note: In qualitative studies, the researcher must disclose all pertinent information that applies to the rationale and conduct of the study. List all preconceived notions about your study topic. List all biases that apply. In this paradigm, researchers articulate their theoretical and practical beliefs. Be sure to tie your choice of methods to your purpose, exploratory questions, and literature.

Later, you will rewrite these first three chapters in the past tense, because the study will have been completed, and you may need to modify the original proposal. Remember, a proposal is a working document and is not meant to be a slavish recipe. Also, analysis is continual in the field, but it is only after leaving the field that the final analysis of data can proceed.

Your fourth chapter should be a presentation of the data. In qualitative work, it may take many forms, such as a narrative, a series of case studies sometimes introduced with found data poetry, a dialogue, a history, or a reader's theater project. You, the researcher, must decide how best to present your data. Use quotations from the interview transcripts in your presentation, and remember that more is more. Likewise, you may find that you need to have more than the traditional five chapters of the dissertation, and that is entirely your choice. Be sure to have a rationale for that decision. *Be sure to use abundant sections from your transcripts.* This will help lead your reader to your model of what occurred in the study as well as to understand your major findings from your categories. Be sure to include a visual schema of all your original categories and then the final, distilled few.

Finally, the fifth chapter should be a discussion of the findings, recommendations for future research, and interpretation of the data. In addition, this is the place to respond to your exploratory questions. This is where you, the researcher, have the opportunity to make sense of your study for the reader. Remember that the data do not speak for themselves. It is most helpful to develop models of what occurred in the study. Those doing multiple cases in a study should do a cross-case analysis here and interpret that analysis for the reader. You must always interpret the data for your reader. At that point, your responsibility as a researcher is complete. You do not have control of the study after that, for readers may make more or less of your data and, in fact, may render a completely different interpretation of the data. Your job is to be as persuasive as possible with the evidence to support your interpretation. This evidence comes from the direct words of your participants via the interview transcripts and field observations. You must be precise in this final chapter so that you bring the reader of your research to accept your explanations, conclusions, and recommendations. After all this, you are ready to write the abstract of your study, which is what everyone will read in *Dissertation Abstracts* worldwide. Be sure to include the purpose of the study, the questions that guide the study, the theoretical framework of the study, a description of data collection procedures, findings, and your major interpretive statements, all within the word limit suggested by *Dissertation Abstracts.* Thus, you may already notice that all components of the dissertation are adequately addressed. Now, let us turn to the attributes that will sustain you during your study.

In Figure 6.2, you can see that Ruth ending up analyzing each of the 12 cases and then did a cross-case analysis of the three main categories, the generalists, the technocrats, and the policy experts. By the time the data were collected and preliminary categories developed, she knew that she was going to have three cases with four persons in each case. You, too, will see that when you begin interviewing, and progress through all interviews,

Figure 6.2   Sample Table of Contents of Ruth Slotnick's (2010) Dissertation

(Continued)

Figure 6.2 (Continued)

Chapter Five: Data Analysis
Cross-Case Analysis
  Policy Proximity
  Perceptions
  Policy Fluency
Policy Experts
  Policy Proximity
    FGPU administrators'
    proximity (Devon and
    Sam)
    GECC administrators'
    proximity (Emerson
    and Leslie)
  Perceptions
    FGPU administrators'
    perceptions (Devon
    and Sam)
    GECC administrators'
    perceptions (Emerson
    and Leslie)
  Policy Fluency
    FGPU administrators'
    fluency (Devon and
    Sam)
    GECC administrators'
    fluency (Emerson and
    Leslie)
  *Summary of the Policy
    Experts*
  *Cross-Case Analysis of the
    Policy Experts*
The Technocrats
  Policy Proximity
    FGPU administrators'
    proximity (Alex and
    Bobby)
    GECC administrators'
    proximity (Tyler and
    Teddy)
  Perceptions
    FGPU administrators'
    perceptions (Alex and
    Bobby)

GECC administrators'
    perceptions (Tyler and
    Teddy)
  Policy Fluency
    FGPU administrators'
    fluency (Alex and
    Bobby)
    GECC administrators'
    fluency (Tyler and
    Teddy)
  *Summary of the
    Technocrats*
  *Cross-Case Analysis of the
    Technocrats*
The Generalists
  Policy Proximity
    FGPU administrators'
    proximity (Chris and
    Pat)
    GECC administrators'
    proximity (Drew and
    Jessie)
  Perceptions
    FGPU administrators'
    perceptions (Chris and
    Pat)
    GECC administrators'
    perceptions (Drew and
    Jessie)
  Policy Fluency
    FGPU administrators'
    fluency (Chris
    and Pat)
    GECC administrators'
    fluency (Drew and
    Jessie)
  *Summary of the
    Generalists*
  *Cross-Case Analysis of the
    Generalists*
Cross-Case
  Analysis Summary
Analysis of the Findings

and after you transcript the interviews, the data have a way of coming alive. It speaks to you, and then you have the responsibility of speaking for that data to all who will read your study. This analytical process is dynamic, exhilarating, and truly exciting.

■ Attributes of the Qualitative Researcher

Notice that many beginning qualitative researchers often remark about the difficulty in doing a qualitative research project. They are often amazed at the significant and in-depth time and energy commitment required to complete and sustain the project. Consequently, a forum is needed to discuss, with a reflective posture, exactly the qualities needed to complete such a project. Over the past 20 years, students have engaged with me in such discussions, and so we offer this list of qualities that may be helpful for new researchers.

Attributes Needed to Conduct and Complete the Project

1. A high tolerance for ambiguity

2. A strong determination to fully complete the study

3. A willingness to change plans and directions as needed

4. Resourcefulness and patience

5. Compassion, passion, and integrity

6. Willingness to commit to time in the field and equal time in analysis

7. Ability to trust and question others

8. Ability to know one's self

9. Authenticity

10. Above-average writing ability

11. Ability to focus and not allow distractions

12. Discipline to write every day

13. Diligence

Although this list is not meant to be exhaustive, these qualities emerge in nearly every discussion of this nature. In fact, above-average writing ability is critical in terms of representing data, analyzing it, and interpreting it. The qualitative researcher must be an excellent writer. Of course, many tools are out there for all qualitative researchers, including many resources on the Internet.

■ Interpretation of Data

Interpreting the data after a presentation of major and minor categories of the findings is a chief responsibility of the researcher. Because qualitative work relies on grounding the theory in the data, researchers usually develop a model of what occurred in the study. This model may be represented in visual terms with drawings and graphs or in verbal terms. Look at this model of a study by Judith Gouwens (1995). She studied two principals in the City of Chicago Schools and was interested in describing and explaining their personal reality in terms of what kind of leaders they were. The variables that affected the two principals' personal realities were the following:

1. The nature, range, and variety of experience of each principal

2. The social interaction dimension of these experiences

3. The emotional response to these experiences

4. Reflexivity and meaning seeking of the individuals

With these four categories, Gouwens compared and contrasted the two cases relying on this model from the data in the study.

Although this is just one example of how interpretation of data can be rendered, the point is that each researcher must interpret the data for the reader. It is the final act of the researcher in any given project. In addition, researchers need to be aware that readers of the research may make entirely different meaning from the data and use the data in a way that may even be at odds with the researcher's own interpretation. This is one of the hazards of research in general, yet the researcher's responsibility to complete a study includes interpretation of the data, whatever it may be used for at a later date.

■ Some Thoughts on Qualitative Researchers Interacting With Institutional Review Boards

Over the past 2 decades, qualitative researchers have been dealing with a number of serious issues when interacting with their respective IRBs. Here, I will describe and explain some of the curious situations that doctoral students and faculty have faced when proposing qualitative research projects. Each of the three examples is framed in the context of values, particularly related to the foundations of qualitative research projects. The three examples are as follows:

1. Informed consent

2. How we use language

3. Coresearchers in a project

In terms of values, the issue of informed consent is at the heart of the matter. Likewise, the issue of how language is used and how data are represented is addressed. Furthermore, the issue of coresearchers in a project uncovers various subtextual ethical issues imposed upon qualitative researchers. These examples uncover more problems than solutions. For example, we simply must be watchful as qualitative researchers in applying for approval from respective IRBs. At the same time, a hopeful message can be gained from qualitative researchers taking an active part in membership on IRBs.

Over the years, I have become increasingly interested in the problems arising for qualitative researchers, particularly doctoral students, when the IRB forms are open for review. In recent years, the types of questions asked of qualitative researchers, the ethical issues raised by these questions, and the burden put upon the shoulders of the researchers involved have raised my interest to the point where I need to write about this and include it in this text. These three examples are only the beginning of a dialogue between qualitative researchers and IRB members. I also encourage qualitative researchers to gain membership on these boards. Recently, I have become an active member of the IRB. This section will describe briefly three examples in dealing with IRBs and the resulting misconceptions uncovered about qualitative research projects. How the problems were resolved will also be discussed. In the process of responding to questions posed by IRB members, it was clear that many IRB members possessed the following qualities:

- Had little knowledge of the procedures, theory, history, or work of qualitative researchers
- Had little interaction personally with qualitative researchers
- Were requiring a standard for qualitative researchers unlike that of other researchers
- Had never thought about including qualitative researchers on the board itself

The examples include the following.

Values: Whose Study Is This, Anyway?

In this example, a series of questions was put to a doctoral student. The questions included the following statements by an IRB board member:

"I don't like the topic of your study. This university would not like this kind of topic."

"I don't think you can use the participants' actual names."

"OK . . . so you have their written consent, and they are adults."

"I won't approve this."

The student was valiant in her defense of her choices and, in fact, eventually prevailed by clarifying for the IRB member her purpose, which he had completely misread. She was studying two former gang members' perceptions about their education and how being in a gang affected their education. The IRB member read the application as a study of gangs. She told him to read more carefully, and that, in fact, he had no right to tell her what her dissertation topic should be. Of course, she herself had spent her lifetime working with gang members and was not to be intimidated by anyone.

She also wanted to use the names of the individuals in the study at their request. IRB members objected, so the solution occurred to me to write down all of the letters in each party's name and scramble them into a new name; the compromise was satisfactory to both parties.

Values: Informed Consent or Subterfuge?

In this example, a student was made to redo her consent form four times because she used the words "I understand no harm will come to me." IRB members, a group of mainly psychologists at that time, cringed at the words "no harm will come to me." The sentence was simply rewritten to the effect that the participant realized the implications of all aspects of the study. Thus, the doctoral student treated the situation as an innovation and decided that learning how quantitative researchers use language in their world would be handy. I surely encourage this posture. The more the qualitative researcher knows about the makeup of the IRB board, the fields of study, and the common use of language in those fields, the better off the researcher will be.

Values: Researcher and Participants Are Coresearchers

In this case, the researcher described how, in effect, the participants and the researchers are coresearchers, and she mentioned this in the consent form. She was asked to eliminate it. Although she argued valiantly for describing the work she was doing as coresearching, she put that as a secondary issue to get her study approved. In the dissertation, she wrote of this at length as an ethical issue.

Following these examples, it is clear that qualitative researchers need to become members of IRBs for obvious and not-so-obvious reasons. In my own case, I raised these issues with the provost of my institution and was immediately put on the IRB; currently, I am an active member. This leads me to the next point, which is about qualitative researchers gaining membership on IRBs to effect change in attitudes and practice. Why should we become members? Because we advance our field, build bridges with those who may not know of our approaches to research, and elevate the level of discourse about research. In effect, we open windows to view many other approaches to research. We also uncover all sorts of ethical issues, because one size does not fit all in qualitative research.

I spoke at the annual meeting of the American Educational Research Association recently on this topic and was amazed by the number of similar cases brought up in the discussion. Likewise, many members have written to me expressing their own stories of being held up by IRB approval over issues that were both serious and substantive, but with a little bit of dialogue, were eventually resolved. Obviously, many cases may never be resolved. But, I want to emphasize here that we must continue to educate IRB members about the theory that guides us and the practice to which we are committed.

■ Summary of Chapter 6

Perhaps the hardest parts for first-time qualitative researchers is the analysis and interpretation and dealing with the IRB. However, this can become nearly effortless through some of the suggestions made in this section, such as the following:

1. Keep a journal, and write every day.

2. Don't be afraid of ethical issues that pop up. Instead, write about those issues. Analyze and interpret them.

3. Rely on your intuition and imagination. Yes, that is something every researcher needs to do. Check out the many books on creativity and great scientists, writers, and artists. All of them trust their intuitions.

4. Get educated about your IRB boards. Learn their language, and follow up as needed. Treat work with the IRB as an innovation, and even get on the board!

5. Find your passion. Do not write someone else's study with data so moldy that even the original collectors of the data forgot where the data were stashed.

6. Become a member of the IRB.

7. Find a peer who will read your work and vice versa.

8. Know yourself.

Obviously, a lot of this is common sense, but if we look to the arts, especially dance and yoga, we can learn a lot about the process. All dancers practice, practice, practice. All students of yoga do as well. I think that, to the extent that you exercise your mind as well as your body, you can only be sharper in your analytical and interpretive skills.

Analysis and interpretation of the data in any given research project must include a clear description of unintended moments in the research; intuitive, informed hunches; ethical concerns and issues; and a serious description of the researcher's role in the entire history of the project. Analysis of data is very much like the dancer's floor exercises. Floor exercises follow stretching exercises in dance warm-ups. After floor exercises, the dancer's next step is performance of a given dance. Interpretation of data by the researcher is like the dancer's act of performance. It can occur only after long-term practice and work. The normal checks and balances system in qualitative research work includes a reasonable, long-term commitment to the research practice at hand and relies on the stability embedded in a long-term activity. Likewise, the study of dance relies on stretching and floor exercises before moving into the realm of performance. The exercises described in this book rely heavily on the arts and humanities for their inspiration. Observation exercises, interview exercises, role of the researcher exercises, writing exercises, evaluation exercises, and discussion of ethical issues exercises all provide an opportunity for learners to stretch. These exercises were designed to allow individuals to stretch from one point to another in a focused pattern of practice to educate and inspire them to become better qualitative researchers. By using activities from the arts, like drawing, photography, and dramatic art, individuals may discover new ways of thinking and opening the mind. By relentless writing activities, such as journal writing, description of beliefs and behaviors, letter writing, and self-evaluation, learners widen their repertoire of research skills. Likewise, physical and mental construction of collages, YaYa boxes, wreaths, quilts, posters, and so on also expand our notions of how we can become sharper at research skills in the field by sharpening our senses. The researcher is the research instrument in qualitative research and must be ready to become physically sharper at observation and interview skills. This is like the dancer who relies on his or her body, which is the instrument with which the story of the dance is told. As Martha Graham put so well, using words to this effect, the body is the instrument through which life is lived and which tells the story of the dance. For those of us who pursue qualitative research questions and design qualitative research studies, I hope that the exercises and resources in this book provide ways to approach developing a stronger body and mind for completing qualitative research projects. In closing, here is one final exercise to integrate all we have been studying together and that integrates your artistic side into the qualitative process.

Exercise 6.1:   ■

Design and Create a Cover for the Researcher Reflective Journal   ■

Goal: Now that you have kept a researcher reflective journal while using this book, design and create a cover that represents you. What kind of research instrument are you? How would you like to represent all that you learned doing these exercises? Use any photographs, products, and artifacts that best describe you as the research instrument, you as the qualitative researcher. Make categories of your journal as needed. Insert all your reflections and any of your completed exercises. Make a dedication in the front of the journal to someone who you think would appreciate this. Make a list at the end of at least three goals for you to work on in the future in terms of developing as a qualitative researcher.

Appendixes ■

A. Sample Letter to Participants

B. Examples of Researcher Reflective Journals

C. Sample Interview Protocols From a Completed Study

D. Qualitative Research Methods: Sample Syllabus

E. Sample Projects From Various Classes

F. Sample Consent Form Recently Accepted by the Institutional Review Board

G. Conducting Qualitative Interviews: Rules of Thumb

H. Sample Sets of Themes and Categories From a Completed Study

I. Electronic Resources

J. Qualitative Dissertation Costs

K. Sample Member Check Form

L. Samples of Interview Transcripts (Edited)

M. The Analytical Process of Coding

Appendix A ■

Sample Letter to Participants ■

Dear _____

I am a doctoral candidate in the Department of Educational Leadership and Organizational Change at Roosevelt University in Chicago, Illinois. I am pursuing my dissertation topic on educational leader perspectives on integrating technology into the undergraduate general education curriculum. The purpose of the study is to describe and explain selected deans', technology directors', and faculty members' perspectives on instructional technology use in the undergraduate general education curriculum. Your participation is requested because of your past and current work as an educational leader at an institution that has integrated technology in the undergraduate general education curriculum.

Participating in the study will require approximately two 1-hour, in-depth interviews. The interviews will, with your permission, be taped and transcribed. To maintain confidentiality, you will not be identified by name on the tape. I or a professional typist will be transcribing the tapes. An outside reader will read the transcriptions of the tape; however, they will be able to identify the technology directors and assistant deans only as technology director A or assistant dean B. The tapes will be kept in a safe in my house. Each participant will be offered a copy of the tape as well as a copy of the transcription. The participants and I will be the only ones with access to the tapes after transcription. Once the tapes are transcribed, a master tape will be made from the originals, and they will be erased. The master tape will remain in my possession and will be destroyed three years after publication of the dissertation.

A comparable amount of time will be required for conducting observations by shadowing you in a variety of situations related to your role as technology director or assistant dean. Interviews and observations will be arranged at the college at your convenience. The tentative schedule calls for one interview in February 2002, one interview in March 2002, and an observation in April 2002.

In addition, you may be asked to share relevant artifacts and documents. Your name, the name of the college, and any other information gathered in this study will remain confidential and will only be used for educational purposes.

I appreciate your thoughtful consideration of my request. I look forward to your participation in the study.

Sincerely,

Carolyn N. Stevenson

Appendix B ■

Examples of Researcher Reflective Journals ■

■ Example 1: Reflections on a Qualitative Research Class

By Ruth Slotnick (2010)

My greatest challenge when doing this project has been my own energy level. Quite frankly, I am exhausted. The writing demands of my dissertation over the past 8 months have zapped my ability to fully engage. The next time I do an observation, I need a little recuperative time between projects. This is a valuable lesson learned. I need to rest, relax, read fiction, read qualitative literature, and refuel. As far as meeting my personal goals of systemizing and organization, I feel that I have incrementally improved. Recording entries in a reflective journal, however, still feels like an afterthought. I am not yet regimented and disciplined enough to write immediately upon completing fieldwork. This practice (or laziness) must change or else the process feels half baked.

Merriam (2009) underscores the painstaking process of honing the researcher's eye and developing a razor-edge focus while systematically recording the descriptive detail required for both participant and nonparticipant observations (p.118). [I feel that my skills to capture complexity have greatly increased over the last week.] The researcher's reflective journal combined with taking copious field notes forms the thick texture and rich dimension that make the researched context come alive. [My field notes and reflective journal have added to these descriptive layers.] Otherwise, Merriam notes, "Data are nothing more than ordinary bits and pieces of information found in the environment" (p. 85). Janesick (2004) argues that the reflective journal should be used in conjunction with the field notes. The journal becomes an introspective crucible or self-evaluation device that allows the qualitative researcher a place to mix together perceptions, intuitions, and hunches not necessarily explored in descriptive data. Together, these basic skills help the qualitative researcher fine-tune their research skill sets and continuously sharpen the tools of the "descriptive inquirer" (p. 2). In developing these skills, Janesick states that a novice researcher must locate his or her own "personal velocity" and methodology that "resonates" with who he or she is (p. 6). Stated differently, aside from knowing self, the qualitative researcher must find a pattern and a speed that naturally allows him or her to fully observe, efficiently collect, deeply analyze, carefully synthesize, and ultimately construct a compelling account of the study. To accomplish this, one must be "persistent and indomitable" (p. 8) if they are to achieve the painstakingly disciplined and rewarding work of the qualitative researcher. However, Janesick is quick to note that remaining open to the data and arriving to it each time anew with no preconceived notions, no matter how seasoned one is to the process, is paramount (p. 2). Finally, she stresses that the health of the researcher is imperative to successfully accomplish the qualitative quest because the researcher is the collection instrument. Trust me when I say that it is wise to heed her call.

October 25, 2008: Writing the Dissertation Proposal

I'm organizing the literature review. Have a goal to really get better at synthesizing and writing. I think I will always be a slow synthesizer and writer. The goal is to be precise but take it easy on myself as my brain is desperately trying to gnaw on thoughts. I am not a better-than-average writer, which is listed as number 10 on Janesick's (2004) scale of necessary attributes needed to conduct and complete a qualitative dissertation. For the most part, I possess the rest of the attributes, including an intense focus, passion, and authenticity, attributes that top the list. Ambiguity and flexibility are words that I have to have more patience with, but I must admit that, 5 years ago, I was neither cognitively disciplined enough nor patient enough to pursue a doctorate. Years of feeling a little stupid by a traumatic event that occurred in grade school (I was held back a grade in elementary school after the school year was well underway, which was humiliating in its own right) and a series of academic stumbling blocks, including difficulty with quickly grasping statistical concepts and an inability to fully express myself in writing, led to its own academic paralysis. It haunts me still when I am confronted by these shadowy demons. So, the attempt to organize does help. The papers, the stacks of books from the library, new books with glossy covers—all are just waiting to be read, contemplated, and synthesized. If my head was not such a sieve, I might just remember all the things I underlined and why I thought it linked to something important in my research. Perhaps this journaling process will allow me to have some cognitive dissonance from my dissertation, giving me the freedom to be completely open in the writing process without the fear that it will stand in judgment of my committee. It is my own place to be as stupid and as smart as I feel, a place to record my ideas.

October 26, 2008

When conducting qualitative research, it is imperative for me to enter into a quiet space where I completely give myself over to the scholarly texts to absorb new ideas while moving back and forth from the big picture to the microview or vice versa. Janesick (2004) calls this donning a "meditative focus" (p. 95), where one's self-awareness is enriched and sharpened by the process of reflective journaling. The researcher's reflective journal becomes a repository for all vestiges of thoughts without worry of being judged. Journal pages are places for poetry, prose, pen and ink drawings, fragments, sketches, or well-developed explanations. Writing reflectively, away from the pages of the dissertation, allows equally for a certain cognitive dissonance that takes the pressure off of writing academically. Oddly enough, the writing becomes so intrinsically linked to what I am gleaning from scholarly readings, philosophical conversations, and observations that a natural synthesizing of the material occurs that can be easily woven back into the dissertation.

Reflection: On Analysis

May 30, 2009

I am in the process of what Piantanida & Garman (1999) describe as "living with the study" and "slogging through" as I gather my research "stuff" (p.129). *Stuff* is described as all things pertaining to the study. Stuff is defined as more than the numeric data. It is anything and everything that goes into collection for data analysis. Stuff feels larger and rounder and perhaps less formal or not so cold as the word *data*. Then, they define five distinct phases or facets (their term) of living with the data:

1. Immersing oneself in inquiry

2. Amassing the stuff of the inquiry

3. Slogging through the stuff

4. Coming to the conceptual leap

5. Crafting portrayals

What exactly do they mean by portrayals? Does that mean my interpretation of what was portrayed via the stuff? Does it mean my descriptions or renderings? Or putting together the picture? Is the crafting of portrayals just like doing a textual reading of Shakespeare? As the director or actor, how would I portray the character I am acting? Isn't the reading of a play a hermeneutic journey in itself? Piantanida and Garman continue describing the analysis process, advising the novice researcher that the last step is "generating knowledge through crafting portrayals" (p. 130). The word *crafting* is really important for it implies that the human hand of the researcher interprets and shapes the text. I need to define crafting a bit more. Crafting is defined as a skilled activity. The text becomes the object to be shaped by the hand of the researcher—the researcher's clay.

The authors also note that the struggle for novice qualitative researchers is grappling with "small, in-depth, context bound inquiry" (p. 131). What legitimates such an inquiry? Distinguishes this critical component compared to quantitative inquiry that prides on generalization when small studies yield smaller more micro results? Rich inquiry, no matter how big or small the interview pool, can lead to a wealth of knowledge inside an academic institution. The aim of the qualitative dissertation in education, the authors' note, is "to generate deeper understandings and insights into complex educational phenomena as they occur within particular contexts" (p. 132). In my study, I am looking for the manifestation of the phenomena I am studying in my specific context. As I craft the portrayal of my study, the integrity and rigor are judged by the depth of analysis that I give to data, the stuff. Even what I am writing here is part of the crafting. I want to write more about my idea of the researcher's clay. The stuff becomes the vessel. Everything goes into stuff. Everything in the studio—lights, electricity, kiln, chemicals, water, clay, potter, and energy—all produce the handcrafted, finished piece.

■ Example 2: Entry From the Technology Study

By Carolyn Stevenson (2010)

The second interview with faculty member B went well. She was surprisingly cheerful, open, and candid. Although this was our second audiotaped interview, I felt I had known her a long time because of the nature of our relaxed conversation. Two elements really stood out as I listened to the tape after the interview: mandatory use of technology and collaboration. During several points of the interview, she mentioned that technology use was mandatory. There was a sense of animosity toward technology use and "hoop jumping." The administration set goals regarding technology use, and they had to be met. I admire [B] for her ability to meet these goals. The other concept was that of collaboration. She mentioned that, with previous projects involving technology integration into the curriculum, peers trained each other. [B] valued collaboration as demonstrated in providing individual training sessions for faculty. She brought up a good point, that often people are intimidated in large sessions or forget some of the information. By working with small groups, she made others more comfortable with the technology use.

She also mentioned that many people were willing to use the technology because she asked them. This statement is also a testimony to her authentic personality. During our interviews, she always had a pleasant demeanor and cheerful approach to discussing even difficult or challenging topics. I also observed that in her interaction with the English department chair, she was cheerful and spoke at the same rate of speed as in her interviews. [B's] rate of speed was conversational, devoid of any nervousness. Her first tape was, well, not easy, but was the easiest to transcribe. She spoke in a clear manner at a moderate rate of speed.

One of the issues that keeps playing in my head is that of the challenges she faced trying to get the entire faculty on board. The amount of motivation was great. I mean, what would happen if the faculty didn't use Prometheus? Would that threaten their jobs? I don't think so. [B] had made reference to certain faculty members who were resistant to technology use, and that was just the way it was. Certain faculty members are more open to technology use while others will never use it. The feeling of "hoop jumping" was mentioned by [B] in the first interview as well. What a difficult job it must have been to try and convince people of the benefits of technology use.

I think that technology can be used to do a job better and faster. However, I guess I am uncomfortable with the notion of almost forcing people to use technology; that might not be the best way to go. This thought is especially true in the area of distance learning. If instructors are not comfortable teaching in an online environment, I believe they should not be forced into teaching in that format. [B] believes that the college will not go the complete distance learning route, and I think that makes sense for this college. In my opinion, just because there is a new technology available or new delivery method for a course doesn't necessarily mean that every college should jump on the bandwagon.

■ **Example 3: Entry Upon
Completing the Interview Project**

By Jason Pepe (2007)

Reflection: Skills, Difficulties, and What I Would Change

This project was both rigorous and relevant, providing me ample opportunity to hone my interview skills. As I begin to narrow my dissertation topic, the interview process gains significance. Thus, I must understand how to utilize the interview process effectively to gather reliable data. In the course of this project, I learned that interviewing someone effectively means much more than just knowing how to write exceptional questions. Certainly, writing thoughtful, open-ended questions is an essential component of the interview process; however, the process does not end there. In fact, the questions are only as insightful as the person asking the questions. For example, my eight-year-old son can read aloud the questions listed in Protocol A and B, but he is unable to conceptualize, theorize, and infer meaning from the person's answers. Without these skills, he is unable to ask probing follow-up questions based on the answers to the original questions. Instead, he politely listens to the person's answer and continues to the next question. A good researcher uses all the senses, even intuition (Janesick, 2004), to see beyond the question-answer cycle, digging deeply beneath the surface to get to the core of the matter being discussed.

At risk of sounding haughty, I believe I am a strong interviewer. Listening is one of my strong suits. I know how to actively listen, and I rely on this skill while conducting interviews. While [S] talked, I carefully listened for connections and contradictions at the same time. Did something she said connect to an earlier reference? Is one of her remarks inconsistent with an earlier one? These answers only come with careful listening and gentle probing. [S] shows us that you can be direct and gentle at the same time without overpowering the person with whom you are talking. My direct approach must be couched in a blanket of genuine sensitivity. If I come across as an interrogator rather than a trustworthy person asking poignant questions, then [S] is likely to shut down or provide guarded answers to my questions. The result is unreliable, or at best, insufficient data for a research study.

Although my listening skills are strong, there are other areas that I need to develop in order to improve my interviewing techniques. For example, the transcript reveals my difficulty with formulating some of the questions. On occasion, I asked a question multiple times until it is presented in the manner I prefer. Merriam (1997/2001) warns against the use of multiple questions. A compound question is confusing and alters the direction of the conversation. Master interviewers ask precise, crisp questions that guide a conversation or elaborate a previous comment. I plan to improve my own questioning techniques with conscious practice. What fascinated me most was the fact that *reading* brought this challenge to my attention. I never reflected on this difficulty while watching the video, but I certainly noticed this problem as I read the transcript.

As I reflect on this project, there are a few parts I would change. I conducted the first interview on February 18, 2009, and the second interview on March 13, 2009. I allowed too much time between sessions. Although some time between interviews provides reflection and preparation time, waiting too long impedes the process. Think of an athlete who must rest her muscles between workouts; if she waits too long, she loses muscle tone, flexibility, and strength and must start over to regain what was lost. A week between Interview 1 and 2 provides enough time to reflect and pull out themes. In addition to an ideal time span between sessions, the setting of an interview is vital to the success of this process. Although [S] and I were never interrupted during either session, I would still change the location to a place off campus. The fact that I could be called on the PA system or by radio at any moment left part of me on duty during the interview and prevented me from devoting my undivided attention to the matter at hand. At the end of the second interview, I even commented to [S] that I felt guilty using school time to complete a project for the university. She disagreed and stated that I was pursuing an advanced degree in education to assist with my professional development. Other professionals, she added, are encouraged (and compensated) when they further their educations during the workday. She makes an important comparison that directly connects to the social status theme outlined above. Once again, [S] models a quiet and elegant strength and masterfully says, "No."

■ Example 4: Reflections Following the Observing in a Student Lounge

By Derya Kulavuz-Onal (2008)

This observation assignment was very enjoyable for me, overall. First, it made me feel like a real qualitative researcher for the first time. I have always been known as a good observer among my friends; however, this was the first time that I consciously and systematically observed a social place and people to discover and explore about that social place. This helped me to really understand what observing for research purposes would be like, what challenges I would face, and how to deal with them.

This observation was a nonparticipant observation (Janesick, 2004), also referred as observation as an "onlooker" (Patton, 2002). I decided to observe this place as I sensed something unique about it from my previous informal observations, which I stated earlier in this report. Therefore, my choice was intentional. Although Janesick (2004) suggested that the observations follow an order (first observation is observing the setting, second is observation of people), I preferred to include as many details as possible when appropriate. Therefore, I tried to be attentive to many things at a time because I believe, although you may focus only one aspect of the assignment in one observation, you can still discover something about that aspect in other observations. This is the reason why observations need to be done more than once, so that, whatever the observer misses in one observation, there is still chance to get it during the other observations. For this reason, I tried to write down everything I could observe during the observations. I also used my laptop, which gave me a faster speed and enabled me to look just like another student rather than an observer. In addition, I used a field note format suggested by Janesick (2004), the aim of which is similar to an observation protocol recommended by Creswell (2007). This format enabled me to organize my observations in a better way.

I think it is important for a nonparticipant observer not to look like a stranger so that nobody pays attention to you. During these observations, I used my laptop to take field notes. This helped me take notes faster and look at the environment without making disturbing head movements. It was easy to observe people over the laptop screen, and it did not catch any attention. As it was also very common for the students to use laptops in this area, I think I looked like a student who was trying to concentrate on her work. Moreover, when I was observing the student organization meeting, I could immediately search on the Internet what this group was about. This was also helpful in the sense that I could make more sense of what I was observing.

The main difficulty I had was that the student lounge was a very active place in the afternoons. Every 5 minutes, the number and the diversity of students were changing. It was not easy to keep track of the demographics of the students in this lounge. Janesick (2004) also expresses that it is more difficult to observe people than in a still scene as people are always active. The other challenge I had was more related with the structure of the setting. The column (the pillar) near to the open area of the first floor always made it difficult to observe a table no matter where I sat. I had to lean side-ward to see who was sitting there when I needed to count the number of students, which might have been awkward and caught attention of the people around me.

If I were to return to the setting, I would follow different strategies in order to overcome the difficulty I had to keep track of the demographic information during my observations. I would draw a detailed seating chart of the lounge for myself and make many copies of it. I would keep them with me during my observations in addition to my field note chart. I would give a number for every table. Every 15 minutes or 30 minutes, depending on the duration of the observation of the time of the day, I would write down the number and diversity of the students on each table on this chart, rather than in my field note chart. Writing the number and diversity of the students in the field notes makes it really confusing, and I think this infor-mation should be separate from the other observations. In my opinion, it is easier to figure out the demographic information if it is in one place.

What I did for this assignment can be categorized into an unstructured observation, which helped me observe whatever is possible to observe through a naturalistic approach (Mulhall, 2003). However, in returning to the setting after these observations, for certain things that I want to con-firm, such as the dominant ethnicity of the setting at certain times of the day, I would use a structured observation approach. These structured obser-vations would still be accompanied by unstructured observations, but they would make it easier for me to confirm data from previous observations and focus more on discovering new aspects of the setting.

Overall, this assignment enabled me to see that observations are not only research tools but also a part of life. I realized that we are observing every-thing around us every day. However, when it is systematic, intentional, and accompanied by note-taking, it becomes more research-oriented. I also strongly believe that whatever type of research a researcher is doing, observing the setting and taking field notes of these observations should be the first step for data collection. Observations equip the researcher with background information of the social context that the research will take place in. I believe understanding the social context of any setting that is subject to research is essential in any type of research, be it qualitative or quantitative. Moreover, by the help of this assignment, I see myself fitting more into a qualitative researcher profile, as I feel I really benefit from observing the settings and contexts and build connections between my research and this social context. However, there are still questions that are evoked by these observations and that stay unanswered. I think an observer needs to report and reflect on an observation before moving to the next observation in order to be more aware of what to observe and see if any other data collection methods are needed. This would be the main approach that I will follow while doing observations for my future studies.

References

Creswell, J. W. (2007). *Qualitative inquiry and research design: Choosing among five approaches*. Thousand Oaks, CA: Sage.

Janesick, V. J. (2004). *Stretching exercises for qualitative researchers*. Thousand Oaks, CA: Sage.

Mulhall, A. (2003). In the field: Notes on observation in qualitative research. *Journal of Advanced Nursing, 41*(3), 306–313.

Patton, M. (2002). *Qualitative research and evaluation methods*. Thousand Oaks, CA: Sage.

■ Example 5: Reflection on the Bean Depot Observation

By Chuck Bradley (2009)

Reflection

When I first decided to observe the Bean Depot Café, I did not anticipate the journey I was about to take. As Janesick noted, quoting one of her dance teachers, "the reason to observe so carefully," she said, was "to become more aware of your own body and mind" and to "internalize" the movement (Janesick, 2004, p. 17). During this observation, I felt that I was internalizing all that the Bean Depot Café represented, including its history, patronage, and purpose. I admit that I was a bit hesitant and nervous upon arrival for my first observation. My first impression was that the café was somewhat redneck, and that I did not fit in. The proprietor met me with a skeptical look, asking, "You're not a cop, are you?" I replied with a grin, "No!" "Fed? IRS?" "No and no!" "Then, you are welcome," he said and shook my hand, smiling. I explained my purpose—to observe three times, once for the environment, once for still life descriptions of people, and once for interactions. I explained that I was doing this for a class. He immediately took interest and mentioned several resources for historical background information. I noticed at that moment that I became a part of this place in a small way. As I sat and began the observation, I couldn't help but think about the people that built the building and those who had entered its doors through the last century. Perhaps some of them were famous, but what impressed me was the real feeling of the place. It was a feeling reflected in the patrons and employees, many with historical roots to the area. They were connected through time to those that came before. I sensed their mission to care for this place, to tell its history, and to protect its future. It was a call that I found myself accepting, even if my part was small.

I also wondered about the people whom I was observing. I noticed that I tended to sort them into categories of locals, snowbirds, and tourists. Even within these categories, I tended to create subcategories. For example, within the locals category, I tended to think of El Jobean locals and regional (Port Charlotte) locals as well as redneck and retiree groups. It surprised me that I labeled people so quickly based on appearances and interactions. In some cases, accents and colloquialisms triggered regional classifications, especially with nonlocals, including Northerner, Midwesterner, New Englander, and so on.

During my first observation, the proprietor came up to my table and said, "If you want to see something really amazing, go take a look out front," and motioned to the other front of the building. As I stepped outside, I noticed a little girl about four years of age. She was sitting on the ground and was hammering a nail into a small piece of wood with a little tack hammer. Her mother, sitting in a porch chair, said, "Isn't she something?" At that moment, I was struck with the thought that she is the future of El Jobean, the next generation that will preserve and care for this piece of history. What will the café mean to her? What challenges will she face as she attempts to preserve this place and pass it on to the next generation? Back at my table, the proprietor stopped by a little later to look at my notes. He was not really interested in what I was writing but commented on how neat they were. I asked if he was looking at my chicken scratches, and he replied, "This is the sign of an organized mind." I commented that I had him fooled, and we shared a laugh. I felt that trust was built a bit more at this point as I tried to maintain an honest and open rapport with him (Rubin & Rubin, 2005).

I returned for my second observation on a very cold day. As the café is primarily an outdoor venue (the indoor space is actually a protected porch), it was closed due to the conditions. I took the opportunity to take outdoor pictures without customers and employees. As I took the pictures, I also visually observed more of the features of the café. There is a little, yellow child's chair in front with a picture of Spongebob Squarepants. I wondered if it belonged to the proprietor's granddaughter, whom I observed previously. I noted the well-worn paths around the building, perhaps as old as the building itself. I walked up on the stage and viewed the area from that vantage point. I was impressed that I could barely see the highway less than 20 yards away. I noticed that I felt a responsibility to be careful not to disturb the setting in any way. It was very quiet with a slight breeze occasionally rattling the palm fronds. I paused for a moment to imagine the building as it might have appeared during the days it served as a Florida frontier grocery and general store. I imagined horse-drawn wagons, the smell of burlap bags filled with various staples, people on the porch sharing outrageous fish tales, the folks who retrieved their mail, and the farmers who traded their fresh fruits and vegetables for other supplies. This place truly served as the nerve center for the emerging town of El Jobean.

I returned again on a warmer day and completed the second observation. Although I previously mentioned that the outdoor customers were aware of my actions, the indoor customers were not. I actually asked the proprietor not to tip them off, as I wanted to practice being an unobtrusive observer. Although I caught eyes with them several times, they seemed oblivious to my presence. It seemed that they were locals sharing fishing stories and other local events with each other. The man at the bar also seemed like a local, friendly with the kitchen staff as he nursed his beer. I heard several questions in my head as I observed. Who are these people in relation to the community? What sort of lives do they live? How are they connected with each other? How are they connected to the history of this place? The couple and child who entered were obviously tourists or out-of-town visitors who stopped in to see the depot. The little boy was quite energetic, running back and forth in the dining area until his mother took his hand. They visited the small museum, looked around a bit, bought a coupled of canned drinks, and left. I also found their visit curious. Who were they, and where were they from? Where were they going? What caused them to stop in at the café? What did they think about it, and what piece of it did they take with them?

My third observation was by far my favorite. Watching the band set up brought back many memories of my days on the road. As they proceeded through the sound check, faces and voices echoed in my head, causing feelings of nostalgia and youthful adventure. I remembered the difficulty of being a road musician, including the process of setting up, tearing down, and packing up. I empathized with thoughts that I imagined these performers might also have—it isn't nearly as glamorous as many may think. I wondered how many strained backs, bruised shins, and stubbed toes they may have had in addition to the occasional more serious injury. I remembered the fatigue of one more show that resulted in short tempers and cross words. It seemed that this band deliberately attempted to maintain relaxed relationships with banter and camaraderie encouraged.

Although I focused on the band in my observation, I also noticed the local patrons enjoying their drinks and snacks. There were a few smokers, but the wind was just enough to remove all but the faintest hint of smoke. It was a festive atmosphere with laughter, crowd noise, and wonderful music. As the band began playing crowd favorites, some would get up and dance. I could feel myself smiling, remembering how my parents used to dance to their favorite songs, embarrassing my sister and me. What also impressed me was the neighborliness of the people. Unlike big cities where people tend to stay to themselves, essentially ignoring the people at other tables around them, the patrons of the Bean Depot Café seemed like one community, talking freely with neighboring tables. Indeed, it seemed as if El Jobean were one big family.

On all three occasions, my wonderful hosts fed me without my asking but certainly with my gratitude. On the first visit, they gave me a generously sized cheeseburger topped with a delicious spicy chili. On the second visit, they gave me freshly fried fish and chips. And on the third, they gave me a sampler of fried fish, fried cheese, and jalapeño poppers—breaded and fried jalapeño peppers, served with a sweet and spicy citrus sauce. The food wasn't fancy, just plain good. Of course, it further enhanced the experience as I finished each observation with a cold, frosty draft beer!

For my part, I attempted to provide a holistic, thick description (Merriam, 1998). However, I noticed much more than I provided in print. I noticed my own emotions and thoughts as I progressed from timid newcomer, to knowledgeable observer, to advocate and friend. I found myself imagining what sort of lives the patrons and employees of the Bean Depot may have. Where did they come from? What did they do? What are their friends and families like? Why are they here? I feel I've experienced the qualitative role of historian (Janesick, 2004).

As I finished my last observation, I again met with the proprietor. It was almost painful to tell him that my work was finished. I was amazed at how much I enjoyed the process and wanted to extend the time as much as possible. He looked me square in the eyes, shook my hand, smiled, and hugged me. "Now, you can just come and party with us!" At that moment, I knew that I had new friends and a place where I was welcome any time. I plan to return often and celebrate the work of those early pioneers that created this wonderful little café in a small corner of paradise known as El Jobean.

After a period of time away from the observations and my notes, I returned to think again about the café. If I had all the time in the world, what three things would I want to study more in depth? First, I realized that, although I mentioned the proprietor, I did not reflect on what he appeared to mean to the café. He seemed to be everywhere all at once. He was very much a people person, talking with staff, greeting customers, and taking time with his granddaughter. The café seemed to revolve around him, and he served as ambassador, host, historian, problem solver, and guardian all at the same time. If I had time, I would definitely spend more time studying his interactions with the employees and customers. Second, I would spend more time exploring the objects found in the café for historical clues. What did the objects mean? Were there indications of their owners and purpose within the community of El Jobean? Why were they selected to be included in the décor of the café? Third, I would spend more time studying my own thoughts and connections as I took time to be in this place. Many of the interactions reminded me of my own family, including similarities in stories, language usage, accents, and laughter. There was something intensely familiar about this place that left me feeling safe, relaxed, and connected to this community. It reminded me very much of my experiences in Key West and other laid-back, semitropical locations. The old building, the irreverent, somewhat quirky sense of humor, and the live-and-let-live attitude that immediately conveyed a value for the uniqueness of each and every person without criticism. It was a place in which every person felt celebrated and valued. Like a favorite pair of worn shoes or faded jeans, it was comfortable and inviting. It was a place in which one could easily step in time to wonderful music and step back in time to a place in Florida's history.

Appendix C ■

Sample Interview Protocols
From a Completed Study ■

By Carol Burg (2010)

■ Interview Protocol 1

1. [Name of doctoral student] has nominated you as a mentor. Can you describe to me how you view your role as a mentor to [name of student?] Or, if permission was not given to use the student's name: Can you describe to me how you view your role as a mentor to doctoral students?

2. When you were a mentor to [name of student], what was the mentoring experience like for you? Or, if permission was not given to use the student's name: Typically, what is the mentoring experience like for you?

   a. Follow-up probing question: Were there any negative experiences for you as a mentor?

3. Keeping in mind your mentoring experiences with [name of student], what type of activities did you typically engage in with [name of student], and what do you consider to be your most effective or important mentoring activities?

   Or, if permission was not given to use the student's name: Think of a specific mentoring relationship that you felt worked well; what type of activities did you typically engage in with doctoral students whom you mentored, and what do you consider to be your most effective or important mentoring activities?

4. What motivated you to engage in these activities with the student [or name of student, if permission is given to use the name]?

5. How did you learn to mentor?

6. Can you describe how you decided to be [name of student's] mentor? Or, if permission was not given to use the student's name: Can you describe how you decided to be someone's mentor?

    a. Follow-up probing question: Are there some general qualities of a protégé that you look for?

    b. Do you have any documents or artifacts from your mentoring relationships that you can share with me?

7. Is there anything else you want to tell me at this time?

■ Interview Protocol 2

1. In revisiting our first interview, is there anything you wish to add to your statements on mentoring?

2. How would you define the term *mentor?*

3. In the ideal, what would help insure excellent mentoring?

4. When you think about your life as a mentor, what can you tell future mentors?

Appendix D ■

Qualitative Research
Methods: Sample Syllabus ■

This is a basic and most current outline for my courses on qualitative research methods. This is inserted here as an illustration of some possibilities for those interested in what a course might look like. I change the texts every other term to keep up with the latest new offerings, which will assist learners and keep us all up to date. I also focus on one major text in each course. In the case study methods course, I focus on a case study text. In every course, I also use at least one book on interviewing and my own text, *"Stretching Exercises."* I then ask students to read and report on one completed book that reports a fully nuanced qualitative study.

Themes:

Some set great value on method, while others pride themselves on dispensing with method. To be without method is deplorable, but to depend on method entirely is worse. You must first learn to observe the rules faithfully; afterward, modify them according to your intelligence and capacity. The end of all method is to seem to have no method.

Lu Chai, The Tao of Painting *(1701)*

Perhaps the greatest of all pedagogical fallacies is the notion that a person learns only the particular thing he is studying at the time.

John Dewey (1967)

■ Example: Qualitative Research Methods

The Frame of the College, The College of Education CAREs

The College of Education is dedicated to the ideals of collaboration, academic excellence, research, and equity and diversity. These are key tenets in the conceptual framework of the College of Education. Competence in these ideals will provide candidates in educator preparation programs with the skills, knowledge, and dispositions to be successful in the schools of today and tomorrow. For more information on the conceptual framework, visit: www.coedu.usf.edu/main/qualityassurance/ncate_visit_info_materials.html

How This Framework Connects to the Current Class

In this course, learners will conduct research with the spirit of excellence in technique, collaboration with participants, and with an awareness of finding the stories and voices of those usually on the margins of the dominant social group.

Course Description

This course introduces the understanding of qualitative case study methods, including but not limited to ethnographic cases, grounded theory cases, oral history and other narrative cases, theory, design, data collection, analysis, and interpretation through the art of the narrative. Participants will analyze and interpret observation and interview data as well as learn case method techniques to analyze documents, archival techniques, and multimethods integration as needed and if appropriate. Use of digital cameras and video recorders will be a central technique studied this semester. Ethical issues in fieldwork and the role of the researcher will be key topics for discussion and writing. Students will identify the theoretical frameworks that guide their work and potentially their dissertation projects. Students will also practice keeping a researcher reflective journal throughout the semester. Members need to know the theoretical frameworks suited to qualitative case methods, identify questions appropriate for case studies, and be able to distinguish bounded case studies as one of the following: iterative intrinsic case studies, historical and observational case studies, narrative cases, or multisite case studies. Students will read and discuss the leading writers in these categories.

Goals and Outcomes

In this course you will be expected to execute the following:

- Complete all readings and in-class assignments.
- Master and deepen library research and archival skills.
- Practice interview and observation skills.
- Identify your theoretical framework, such as phenomenology, critical theory, interpretive interactionism, chaos theory, and so on. Describe it, explain it, and use it.
- Identify which of the qualitative case approaches suits you, such as ethnographic cases, case studies, autoethnography, oral history, life history, portraiture, and so on, within the context of bounded case studies.
- Practice writing analytical vignettes based on fieldwork.
- Discuss in small groups and groups as a whole problems related to fieldwork, including ethical issues that arise in research.
- Continue the study of major writers in the field and learn how to read research articles as a reflective, critical, active agent.
- Understand the knowledge base underlying various approaches to research.
- Understand which questions are suited to qualitative case techniques.
- Practice using digital tape recorders, video recorders, and digital photography techniques as data collection tools.
- Understand how the review of literature is essential to the methodology portion of the dissertation and how they are interrelated.
- Demonstrate proficiency in use of web-based resources, blogs, virtual interviews, and use of documentary evidence.
- Understand the differences in the language of case methodology.
- Read, write, discuss, and reflect upon key issues in qualitative case research methods.
- Study issues of race, class, and gender, practice—like locating oneself ideologically—purposeful sampling, and ethics in fieldwork.

Instructional Methods

You will be assigned small-group discussion and group work, computer lab and library work, and projects in and out of the classroom. You will experience some guest speakers (to be announced); some lecture, demonstration, and discussion; Socratic methods of questioning; and minilectures as needed. Documentary film techniques and transcription techniques will also be covered.

Equipment Needed

It is helpful to have a digital voice recorder for interviews. Although a minicassette works, most transcribers prefer a digital voice recorder with a thumb drive insert, such as the Olympus 1100 or higher, any MP3 player, or iPod with appropriate attachments. The cost of a digital recorder is anywhere from $50 up to $150. Many students buy their equipment at major discount stores. Make life easier for yourself by using a digital voice recorder. In the event you do a qualitative dissertation and need transcriptions done, today's transcribers prefer and charge less for digital interviews that can be sent to the transcriber via e-mail.

It is optional, but a digital camera or a combo digital camera and video recorder are also quite helpful. The most user friendly and affordable is the Flip Ultra 60-minute camcorder with built-in USB. This records up to 60 minutes of an interview (for those who decide to do video interviews). The cost is around $150. This type unit also plugs into a TV monitor so you can see your interview or observation at any time in class or at home. Note that 8 megapixels is highly recommended. Another option is the Aiptek mini video recorder, which is the $100 to $130 price range. Check the megapixels and the ability to hook up to your TV. Many prefer the iPod nano with video camera capability. The 16-GB nano is about $169 at a university bookstore or computer store.

Those doing their own transcripts might consider a foot pedal system. Olympus makes a device for about $120. You need to weigh that against the going rate for transcriptions in the open market. Currently, a 1-hour interview transcribed from a digital voice recorder is around $110. If you use a minicassette recorder, the cost is higher.

Professional Transcriptions

Check transcription rates at the top transcription center in the United States, located in Los Angeles, at the following websites:

www.productiontranscripts.com

www.castingwords.com

For up-to-date blogs, narratives, advice, and all things qualitative or otherwise to help you reach your goal of finishing the dissertation, go to the following web addresses:

www.academicladder.com

www.dissertationdoctor.com

These sites include rules of thumb for each chapter of your dissertation, support groups for getting through the process, and blogs.

Local transcription services and copy editors are also available. These will be announced in class.

Required Reading in Common

Now is the time to build your methodology library.

Merriam, S. (2009). *Qualitative research: A guide to design and implementation* (Rev. ed.). San Francisco: Jossey-Bass. (I use this as a basic text in the case methods course.)

OR

Creswell, J. (2007). *Qualitative inquiry and research design: Choosing among five approaches* (2nd ed.). Thousand Oaks, CA: Sage. (I use this in the Qualitative Methods I course.)

OR

Berg, B. (2007). *Qualitative research methods for the social sciences* (6th ed.). New York: Pearson. (I use this in Qualitative Methods II as I have members from medicine, nursing, sociology, communications, business, and psychology in the course in addition to education students.)

THEN

Janesick, V. J. (2004). *"Stretching" exercises for qualitative researchers* (2nd ed.). Thousand Oaks, CA: Sage.

Rubin, H. J., & Rubin, I. S. (2005). *Qualitative interviewing: The art of hearing data* (2nd ed.). Thousand Oaks, CA: Sage.

Select one of these optional texts for your reading, reporting to the class, and future use. These are completed studies or methods supplements:

Gawande, A. (2007). *Better: A surgeon's notes on performance.* New York: Metropolitan Books.

Beavan, C. (2009). *No impact man.* New York: Collier.

Kingsolver, B. (2004). *Animal, vegetable, miracle.* New York: HarperCollins.

Goldberg, N. (2005). *Writing down the bones.* Boston: Shambala Press.

Sacks, O. (1995). *An anthropologist on Mars.* New York: Vintage Books.

Reinharz, S. (1992). *Feminist research methods in social research.* New York: Oxford University Press.

Ehrenreich, B. (2001). *Nickel and dimed: On not getting by in America.* New York: Metropolitan/Owl Books.

If you know you are going to do a qualitative dissertation, you need this text:

Piantinida, M., & Gorman, N. (2009). *The qualitative dissertation.* (2nd ed.). Thousand Oaks, CA: Corwin.

Check the blackboard daily!

Electronic Resources

Electronic and other resources for qualitative researchers include journals, websites, LISTSERVs, software, and so on.

All class members go to http://www.aera.net/Default.aspx?id=777 for a list of relevant sites.

Also check the qualitative methods blogging if you like.

There are key journals in hard text and online journals that focus on qualitative inquiry. There are numerous web resources and transcription services. See pages 124 to 142 in Janesick's *"Stretching Exercises" for Qualitative Researchers,* second edition.

Check out the e-journal *The Qualitative Report* online at http://www.nova .edu/ssss/QR/index.html, and also sign up for their weekly newsletter. *Required!*

Assignments

There will be two major written assignments:

Nonparticipant observation assignment, due on the date of the midterm meeting. (This is 25% of your grade.)

Interview project assignment, due the final class. (This is 50% of your grade.)

Attendance, class participation, technology demonstration, and fieldwork are 25% of your grade.

There will be six in-class assignments, including observation exercises and interview exercises with pass or fail grading.

Each class member must pass the following tasks at the proficiency level.

A. Participant observation exercises, including field notes, thick description, preliminary informed hunches, and what would you do next for:

1. Observation of a still life scene OR

2. Observation of a setting

3. Observation of a person OR

4. Observation of two or more persons interacting

B. Interview, transcription, and preliminary data analysis of same for:

1. Interview of a person you know

2. Interview of a stranger

Transcriptions of at least 30 minutes of taped interview with field notes, preliminary categories for follow-up, and set of informed hunches.

You may use your interview project to fulfill this requirement.

C. Practice with digital photography as needed, and perform in-class demonstrations of technology equipment for the qualitative researcher

D. For the self-selected book that describes a completed case study, each student will report on that text and prepare a 1-page handout for all class members detailing what the book was about and what you learned about case methods

E. Practice with the writing of a researcher reflective journal

Grading

1. Students must inform the instructor before the due date if an assignment will not be ready on time. Students who are absent must find someone to take notes and pick up any in-class handouts for them. Students who miss a class must hand in a 5- to 7-page reflection on the readings for the class they missed. For Saturday classes, recall that 1 Saturday equals 3 weeks or 3 classes of semester time.

2. Students must inform the instructor before class when and if they will be late or absent.

3. If a student is late with a written assignment, a penalty will be included of a flat 10% off your given grade. If a student does not hand in an assignment, the student cannot earn a grade higher than a B.

Projects are worth the following.

Nonparticipant observation assignment: 25%

Interview assignment: 50%

In-class participation, technology use, and attendance: 25%

4. Grading is based on the following scale:

% Mastery	Grade
100–95	A
94–90	A–
89–86	B+
85–80	B
79–70	C

Topics and Schedule: Subject to Change

We will discuss specific pages and chapters for assignments as needed. Pace yourself in your reading time and completion. Build yourself a good library of methods texts and articles.

Class Meeting and Topic	Assignment
1. Intro to class, each other, syllabus, in-class observations	Begin Text 1
2. Questions for research, in-class observation activity	Continue Text 1
3. Nonpart. obs/obs./interviewing, documents, library strategies, obs. activity continued	Complete Text 1; begin Text 2
4. Analyzing interview data, category development, first assignment due, interview activity, in-class narrative methods	Continue Text 2
5–9. Document analysis and archival techniques, computer-assisted data analysis, Interview Activity 2; find your theoretical frame and locate yourself	Begin Interview text
10–15. Qualities of the qualitative, second assignment due, multimethod descriptive validity	Complete Interview text
	Continue next readings Complete all readings

Assignment One: Nonparticipant Observation

Restaurant, coffee shop, shopping mall, zoo, place of worship, museum, health club, funeral parlor, skating rink, book store, library or any public setting.

Why:

To observe a complex public setting. There should be natural public access to the setting and multiple viewing opportunities for you. Due at midterm.

How:

Go to this social setting more than once to get a sense of its complexity and to maximize what you learn. Go at least three times at different times of the day. If you wish to return at any other time, of course, feel free to do so.

Take notes. Make a floor plan. See what you are able to hear, see, and learn just by observing.

1. *The Setting:* Look around you, and describe the entire physical space. Draw a floor plan or take a photo if permitted.

2. *The People:* Look around you, and describe the people in this setting. Focus on one or two of the people. What are they doing in that social space?

3. *The Action:* What are the relationships between people and groups? Try to discover something about the people in the setting.

4. Describe the groups and any common characteristics, for example, age, gender, dress code, speech, activity, and so on.

5. Focus on one person in your viewing area to describe in detail, for example, a waitress, a caretaker, a salesperson, and so on, depending on your setting. Describe your location.

6. If you had all the time in the world to do a study here, what three things would you look for upon returning to the setting? Develop at least three categories or codes.

7. Be sure to give a title to your report that captures your observational study.

8. Be sure to use references from our texts this term or any other appropriate texts.

You have 5 to 7 weeks to complete this assignment. Be sure to include a self-evaluation and discuss the following:

1. Of all the exercises so far, how has this one challenged you? Describe yourself as an observer. What did you actually see?

2. How did you approach this assignment? Why did you select this place?

3. What difficulties did you encounter in the field setting?

4. What would you do differently if you were to return?

5. What did you learn about yourself as a researcher?

6. What three things did you turn into categories or codes?

Assignment 2: Interview Project

This assignment is due at the final class.

Reminder:

This project cannot be handed in any later. No exceptions.

Project:

Interview one person twice, so that you have the experience of going back for an interview. Interviews should be at least 1 hour in length each. The topic of the interview is:

What does your work mean to you?

Select an educator or other professional to interview about what work means to that person. The first interview should have some basic grand-tour questions like the following:

Interview Protocol: A First Interview

1. What does your work mean to you?

2. Talk about a typical day at work. What does it look like?

3. What do you like about your job?

4. What do you dislike about your job?

5. Where do you see yourself in 5 to 10 years?

6. Is there anything else you wish to tell me at this time?

You create the questions for Interview Protocol B based on what you find in the first interview to get to the goal of the interview.

This should take at least 1 hour but no longer than 90 minutes per interview. Aim for each interview to take 1 hour of time. Be sure to take field notes so you can probe into areas of this first interview during the second interview. Be sure to tape all this. Be sure to get informed consent. See Appendix F in this text for a sample.

Due on Last Class Night

A paper or report of 25 to 35 pages that includes at least the following:

a. Describe in detail why you selected this person. Add photo if needed.

b. Provide a list of all the questions asked in each interview and label them Interview Protocol A and Interview Protocol B.

c. Summarize the responses from both interviews in some meaningful way with precise quotations from the interviews.

d. Pull out at least three themes from the interviews.

e. Make a visual model of these themes, and explain your model.

f. Tell the story of what this person's life work means to this person.

g. Include the signed consent form.

h. Discuss any ethical issues that may have come up.

i. Include a sample of three pages of your best transcript from the tape.

j. Finally, include two or more pages of your own reflections on your skills as an interviewer and as a researcher and what you learned from this project; be sure to mention any difficulties that came up and what you would change the next time you conduct an interview. Describe yourself as the research instrument.

k. Be sure to use references from our texts this term (minimum of 7 to 10).

l. Be sure to create a title that captures your themes.

Remember that this is a narrative research paradigm, so you should write this in narrative form as if you were telling this person's story. The story is about the person's life work. See excellent examples on our class blackboard site. Extra credit will be given for integration of photography, video, CD or DVD work, or any attempt to use computer-assisted data coding.

Evaluation of Your Writing Rubric

Use the rubric as a tool to understand how your work will be graded.

Writing Rubric

How Your Writing Will Be Evaluated

	Exceeds the Standard	*Meets the Standards*	*Is Below Standard; Rewrite*

1. _____   _____   _____

The title captures the meaning of the work and is reflective of a theme. The purpose of the interviews is clearly stated. The theoretical frame that guides your method is clearly described.

2. _____   _____   _____

An introduction to the piece adequately captures meaning, content, and the data. A summary and conclusions section is adequately developed and captures the meaning of the interviews and their implications.

3. _____   _____   _____

Adequate data and content are presented, and all questions are adequately addressed.

4. _____   _____   _____

Writing contains references to and understanding of any or all of the texts in class and contains correct grammar, spelling, punctuation, sentence construction, and paragraph construction. Note that all paragraphs must have a minimum of three sentences.

5. _____   _____   _____

Writing is submitted typed, double-spaced, and APA style with appropriate support data, such as references (7 to 10), included as needed.

6. _____   _____   _____

For the interview project, at least 7 to 10 pages of transcript are included in the report.

For the observation project, a photograph or drawing of the physical space is included as well as all field notes.

■ Plagiarism

Copying someone else's work is a very serious offense and can bring about a student's removal from the program and the university. You plagiarize when, intentionally or not, you use someone else's words or ideas without giving them credit. Quotation marks should be used whenever you are using the exact words of another author. Square brackets and ellipses should be used to indicate any words that are deleted from the original material. Summarizing a passage from another source or rearranging the order of a sentence or sentences is paraphrasing. Every time you paraphrase the work of another author, you should give credit to the author by citing. If you are using someone else's ideas, you must give them credit as early as possible in your text. In this course, if you are found guilty of plagiarism, you may receive an F (failure) in this course and be dropped from the program.

Appendix E ■

Sample Projects
From Various Classes ■

This is a collection of sample topics from qualitative methods classes that students have undertaken.

- Looming Large: Linking Lives and Literacy at the Library
- An Interview Study of Florida's Outstanding Teacher of the Year
- An Interview Study of an Emergency Room Nurse
- Traditional Cuts: Life in a Barber Shop
- Parker's Books: A Reader's Delight in Sarasota, Florida
- The Ice House Pub: Where People Go to Chill
- Drink: A Study of a Nightclub and Drinking
- Skin Deep: The Waiting Room of a Dermatology Clinic
- Sideline Sidebars on the Soccer Field
- A Study of a Local Tattoo Parlor and Its Patrons
- An Oral History Study of a Female University Administrator
- An Oral History of a Holocaust Survivor
- Strong Woman: A Study of an Activist Grandmother
- Peace With Bombs: A Study of Military Views on War

Appendix F ■

Sample Consent Form
Recently Accepted by the
Institutional Review Board ■

This study involves interviewing _____ about
_____ and is therefore research.

1. The purpose of this study is _____

2. The study is expected to last from _____
   to _____

3. The number of people to be interviewed is _____

4. The procedure of the research involves asking participants about their
   views on _____

5. The interviews will be 1 hour each in length, and each participant will
   be interviewed twice. The audiotapes will be protected in my home and
   will be kept for 2 years.

6. There are no foreseeable risks to the participants, and they may leave
   the study at any time.

7. Possible benefits are educational, that is, to contribute to the body of
   knowledge about _____

8. Members may choose to be completely anonymous, and all names will
   be changed for reasons of confidentiality. This information will only be
   known to me and the chair of my dissertation committee.

9. For questions about the research contact me _____
   at _____

10. Participation in this study is totally voluntary. Refusal to participate will not result in penalty or loss of benefits.

11. There is no cost to you to participate in the study.

12. The [name your university] Institutional Review Board, IRB, may be contacted at _____ This IRB may request to see my research records of the study.

I, _____ [Print your name here] agree to participate in this study with _____ .
I realize this information will be used for educational purposes. I understand I may withdraw at any time. I understand the intent of this study.

Signed _____

Date _____

Appendix G ■

Conducting Qualitative Interviews: Rules of Thumb ■

1. Begin with a topic, and write a few big questions—basic, descriptive, grand-tour questions.

2. Establish rapport with the interviewee.
   - Establish a relaxed and open atmosphere and pace for the interview.
   - Explain clearly why you are conducting the interview.
   - Tell the interviewee about yourself.
   - Show genuine interest in the interviewee with both verbal and body language.
   - Take notes if possible.
   - Smile.

3. Ask one of the big questions. Based on the response of the interviewee to the big question, ask additional questions to probe and to elicit more complete information. Develop your grand-tour questions for big, empty, time spaces. Revert to statements made earlier if you are worried about spaces in the interview. Relax.

4. Use both body language and verbal language to keep the interviewee talking or to let the interviewee know that you have enough information.

5. Always end in an interview with the question, "Is there anything else you would like to add at this time?"

Appendix H ∎

Sample Sets of Themes and Categories From a Completed Study ∎

By Carolyn N. Stevenson (2002)

∎ Preliminary Themes and Subthemes

Theme 1: Technology Functioned as a Source of Institutional Change.

Subtheme 1: Colleges dramatically changed their images.

Subtheme 2: General education curriculums and departments were restructured and reorganized to integrate technology.

Subtheme 3: Growth was rapid and painful.

Theme 2: Integrating Technology Into the Curriculum Involved Motivating Students, Faculty, and Staff.

Subtheme 1: Students often scared or reluctant to use technology.

Subtheme 2: Faculty required to use technology.

Subtheme 3: Staff required to use technology as positions were transformed.

Subtheme 4: Training essential for faculty and staff motivation.

Theme 3: Implementing Change Involved Leaders on All Levels.

Subtheme 1: Educational leaders enlisted others to promote technology use.

Subtheme 2: Curriculums were redesigned to include technology to assist faculty and students.

Subtheme 3: Technology use focused on practicality.

Theme 4: Technology Use Will Continue to Grow and Expand.

Subtheme 1: Students will be more knowledgeable with technology, promoting increased technology use.

Subtheme 2: Quality of the curriculum should not be compromised by technology use.

Subtheme 3: Distance learning is effective for certain learning communities but not appropriate for all.

■ William Wallace College

The Case of Technological Growth included the following major themes and subthemes:

1. Growth and change within the college fostered technology growth. This institutional growth and change are represented in four subthemes: (a) the developmental steps for integrating technology into the curriculum, (b) change involved motivation and collaboration, (c) leadership needed for change, and (d) continued growth seen in the future of technology.

2. Benefits and challenges of technology. Technology poses both benefits and challenges that must be assessed by educational leaders. These benefits and challenges are expressed in three subthemes: (a) faculty resistance, (b) use of technology in the curriculum, and (c) distance learning is not for every institution.

■ Jane Byrne College

The Case of Transformation Via Technology included the following themes and subthemes:

1. Technology transformed the department and the curriculum. This transformation is represented in three subthemes: (a) the developmental steps for integrating technology into the department and curriculum, (b) motivating others and teamwork needed for change, and (c) leadership as a change agent.

2. Benefits and challenges of technology. These benefits and challenges of technology are expressed in three subthemes: (a) accessibility, (b) technology as learning partner, and (c) distance learning is not for every student.

Appendix I ■

Electronic Resources ■

■ LISTSERVs, Websites, Journals, and Software

Luckily, we now live in the electronic age with resources at our fingertips at the touch of a keypad. Qualitative researchers are able to access quite a bit of information about techniques, design, issues, and problems. Qualitative research LISTSERVs, discussion groups, message boards, software tools, and electronic and regular journals are numerous and easily accessible to all. In fact, the sites are so numerous that listed here are only the top regularly visited sites. A search of the World Wide Web, WWW, and Web2.0 sites will result in thousands of results.

LISTSERVs:

1. www.listserv.uga.edu
2. Actionresearch.altech.org
3. Phenomenologyonline.com
4. Groundedtheory.com

Helpful websites for writing qualitative dissertations, proposals for funded research, and to help new professors in the field:

1. www.academicladder.com

This site provides networking options and writing services for new professors and teachers new to qualitative methods.

2. www.dissertationdoctor.com

This site helps students with brainstorming for topics for study, demonstrates how to do literature reviews, and allows for blogging with peers, along with many other positive resources.

3. International Institute for Qualitative Methodology

www.ualberta.ca/iiqm

4. International Congress on Qualitative Inquiry

www.icqi.org

Journals dedicated to publishing qualitative research:

1. *The International Journal of Qualitative Studies in Education* (QSE)

2. *Qualitative Inquiry* (QI)

3. *The Qualitative Report* (TQR online)

4. *International Review of Qualitative Research*

Software:

Literally, nearly every year, new software packages become available or an existing package is improved. The list below includes some of the top-selling software packages that students have found useful.

1. Atlas-Ti

2. The Ethnograph

3. NVivo

4. MAXQDA

Appendix J ▪

Qualitative Dissertation Costs ▪

Based on the completed study on mentoring by Carol Burg (2010).

Note: If you type your own transcripts, costs are greatly reduced.

Atlas-Ti software	$150
Digital voice recorder	50
Copying costs, paper, postage, notebook	360
Transcriptions of 12 interviews	1250
Starbucks $10 gift certificates (12)	120
Manuscript processing fee	100
Travel to and from sites	150
Microfilm fee	65
Copy editing and UMI fee	465
Final dissertation copies	700
Total (excludes dissertation hours):	$3410

Appendix K ■

Sample Member Check Form ■

March 2009

Dear _____:

Thank you for the insightful and powerful interview(s). Attached please find a draft of the transcripts for your review. Please check for accuracy and that your responses are being reported correctly. Please feel free to contact me at _____ or via e-mail at _____ should you have any questions.

By your act of reading the transcript(s), if I do not hear from you within 5 working days, I will assume you are in agreement with the transcript(s).

Sincerely,

[Your Name]

Appendix L ■

Samples of Interview
Transcripts (Edited) ■

By Carolyn N. Stevenson (2002)

■ Dean B: Interview 2

Date: 3/18/02

Time: 9:30 a.m.

Location: Office of Dean B, Eighth Floor, William Wallace College

Dean B: How has implementing technology into the curriculum affected it? I don't know. The technology was sort of a factor before I became dean. And I am sort of a gadget guy. I think that we probably rely too much on electronic communication. I used to like to go around, and take a tour, and talk to everybody every day, stop at people's desks and chat. And I am reaching a point where I can't even do that.

I spend too much time sending people CC: Mails and CCing people on CC: Mail. I get 86 to 100 CC: Mails a day. And at least a third of them are useless. So, it takes me at least an hour coming in the morning to go through my e-mail, CC: Mail, and phone messages. And that's before my day really starts. Like today, I got here at 7:20 a.m., and it was 8:40 a.m. before I started what I considered to be my first task of the day.

Now, in terms of my teaching, each level of technology has made teaching, for me, easier. I use film a lot. The DVD allows me to cut directly to the scene instead of playing around. Like for example, when teaching *Hamlet,* because there is so much jumping around, I eventually dubbed myself a tape where I had the scenes I wanted all on one tape in the order I wanted them, so I didn't have to jump around.

Where with the DVD, well, let's stop and look at that and then go back. I think that is great, and I think we haven't even touched on the other things, for example, making specific course DVDs for our class, that kind of thing. We are just starting to do that in terms of CD burning, where I can give you a CD, and this has all of my handouts for the quarter. It has the syllabus, the schedule, Internet resources, links, and things, and now, I don't have to give out any handouts.

In terms of Prometheus, that's very handy for the faculty who delve out, and use it, see how it can be a time-saver, a resource for the student. And in talking to people from other campuses, we are pretty far along using courseware than other colleges. Allyn & Bacon want us to be the Prometheus model site. And they told us they were hard-pressed to find another school using Prometheus that has faculty take it to the level where it is being used more than an e-mail portal and a syllabus resting place. I sort of got off track with Question 1.

CNS: That's fine. Keep going if you like.

Dean B: Well, that about covers it for Question 1. Question 2. What messages would you like to give faculty and students about technology use in the curriculum? Well, I discover that students get hung up on, "well, I'm going to make my paper beautiful," instead of, "what is my paper about?" The materials. On the one hand, you have the students on that end of the scale saying, "Well, I can make that look great using technology." On the other end, you have the faculty who don't use it to make anything look good enough. They just think, "I'll just shove my text file into," and they don't think about how they can insert images, that sort of thing.

So, there needs to be a middle ground. When I came here 13 or 14 years ago, I really hadn't even used computers at all. And I think I'm a pretty advanced user. And it certainly makes my job and my teaching better and easier. And I would hope that faculty would take that sort of thing to the next level. I think that gen ed faculty as a group does. Now, within that, there is a continuum, where you have people like [one instructor] on one end who's got things he's been developing for years with technology. And then, there's [another instructor] who does e-mail. So, you have this sort of continuum.

But overall, I think we're in the middle or skewed toward the high end of technology. People are at least using Prometheus as more than an e-mail portal for the most part.

CNS: It is a great tool, but it depends on how and the manner in which it is utilized. Even now, the evening students are getting involved with it.

Students are beginning to anticipate it. And my class, where it is a directed learning class, it is absolutely imperative.

Tell me about your interests in technology.

Dean B: Well, like I've said, I'm sort of a gadget guy. I don't know how I got along without my Palm. I communicate with my daughter in college. Every day in the morning when I sign on my e-mail, she's got it up and running. While she's getting ready for class, we instant message back and forth. I have a computer at home, and I don't do much work at home now. But, I play in a computer baseball simulation. They have teams picked from all of the players in history. The computer has a simulation of a season over a course of a regular season. That's kind of a fun thing. I have a pretty awesome team assembled, but of course, everybody's team is awesome. [Laughter.] And it's interesting. You can simulate the whole game in like 20 seconds. So, that's one of my kinds of relaxing hobbies now. Instead of sitting on the couch and vegging in front of the TV, I'll veg in front of the computer.

So now, everything is electronic. I have this baseball encyclopedia, which is like this thick. And it weighs like 100 pounds; now I have a CD that has that on it. It's just easier to work with.

Technology just seems to me to be a way to do things that you would do otherwise faster and better. And that's how I think of technology. Technology should not be an end to itself. I mean, you go back to the car. The car is just a better horse. And that's what technology should be. It should be an end of itself. Well, I'm going to use the computer to do something I want to do anyway . . . create documents or search documents. It's as simple as instead of passing a book around the classroom for everyone to look at the pictures; now you put it on a document camera, and everyone can see it at once. That's the direction.

So, I think I have also talked about four, too. Was the decision to integrate technology into the curriculum well received by other members of the college? I think that's been a while ago now so. . . . When I hire new faculty, and they come and look at the classroom, you hear, "Well, gee, at Northwestern, all we had was a chalkboard." Or blah, blah, blah. So, I think that is something that gets them excited.

I think at first people were a little resistant to the idea because they saw it was some sort of cure-all, or we have to do it just to do it. And now, I think people see it as a tool for teaching better and better involvement by the students. So, I think we have a pretty good view on it. The problem is for the faculty members. How do you find time to create presentations? And also, once I've got a presentation made, does that ossify what I do? How often do I add to it? Or gee, now that I have my presentation on photosynthesis with this neat animation I made, do I ever update that?

■ Vice President, Information Services Technology Director B: Interview 3

Date: 4/19/02

Time: 11:00 a.m.

Location: Office of Technology Director B, Fourth Floor, William Wallace College

CNS:	Let's begin by talking about your leadership. How would you describe your leadership style?
Technology Director B:	I'm trying to think about how to phrase it. I try to keep abreast with things that are happening. Um, I'm nervous now for some reason. Okay, leadership. I try and keep up with new things that are happening. I try and am visionary on what's coming up. I do that by reading articles, being a member of the CIO magazine, keeping abreast. I am a member of several organizations. I believe that knowing what's coming is important, thinking about how we can incorporate it if it is part of your business strategy to bring the technology into the college.

And the other 80% of the time, I am doing managing of projects.

CNS:

With that in mind, can you ever call a day typical? What is a typical day?

Technology Director B:

Not really. It's all about being able to be flexible and being able to keep up with the changes. It seems like, in the technology area, there are always problems. But there are problems because they are always introducing new ideas. They don't phrase it that way, but I interpret that anybody who comes to my door is looking at the business side. They are either looking at ways they can improve services for our students, faculty, or staff, or they are trying to analyze what we have done in the past, so I think about the things we have done in the past and pull that out. Use that data to help them accomplish whatever it is they are trying to accomplish.

So, there really isn't a typical day because it is always changing, but the fact that the busier I am, I know people are thinking about the future things, even though they don't come to me and phrase it that way.

CNS:

What are some of the benefits and drawbacks of technology?

Technology Director B:

From the administrative side, the benefit is, of course, cost reduction, hopefully, because we're hoping technology will improve services for providing better information and costing us less to do that. We hope that the computers are working harder for us . . . the back-end sector.

The weakness, though, is that if the systems are down, everybody needs to go home because they don't know what to do, so there's a concern there that how much do we let the computers do and how much do we let the people do? It is always a struggle I have with our auditors. I feel our internal auditors feel that computers do all the back-end stuff, and we'll do the inputting.

Then there is turnover, which I think is true in all industries right now. And trying to keep everyone trained.

So, technology has good and bad points.

CNS: Does your office handle the training of the other student services departments?

Technology Director B: No, for example, we would handle the systems that would help the financial aid office. We look at the ways financial aid does their work. So, I am very involved with the managing of the people. But the training portion is left within their own department.

We are talking about in the future having a training department. I'm not really sure what that means. It is supposed to be for the new system. The new system will cover admissions, financial aid, placement, and student records, so that the question becomes who trains the workers? But that would be a whole new initiative.

But, there are thoughts right now of hiring a trainer, but I'm not sure what they are going to train them on. You have to know each business practice.

CNS: That sounds like a really demanding role.

Technology Director B: But it's exciting. The history of the college . . . we're looking for a new system, so it's a good time to be in IT because we're going to require the admissions area we're looking at to be looking at the business aspects and see how these offices are talking to each other, looking at the integration of how they work together. So, it's a good time to be in this area. You're on the cutting edge of [William Wallace] because we're going to be changing. If you are involved with those changes, and you're watching it, you're going to learn all about the development.

Because I was here when they had no technology in the department. And I was there with Mike and developing it. That's how I have a lot of insight. It's not that I am better than anybody else, but I happened to be at the right place at the right time and grew with the college.

■ Faculty Member B: Interview 2

Date: 2/27/02

Time: 1:00 p.m.

Location: William Wallace College, Third Floor Lounge

CNS:	What are the messages you would like to give faculty, staff, and students about technology use in the curriculum?
Faculty Member B:	I think that the most important thing with technology is that schools feel obligated that they still keep abreast with the technological changes and that they really feel they owe it to their students to keep them abreast also. You know, you can't stay on an island and teach nonapplied subjects like English and pretend that technological changes aren't happening.
	So, the message I would give to faculty is that it is not another hoop to go through. It is not unnecessary work but that the school owes it to everyone [to keep] abreast of changes. And we are in a postindustrial society; this is a must. It doesn't always tend to be a benefit, but sometimes it does turn out to be more trouble than it's worth.
	But, we are obligated, because of being in the field of education, to explore and learn whether or not it ends up being beneficial. For staff, in terms of even the Management Institute, they have decided to put their entire yearlong course on Prometheus and to teach their protégés through Prometheus. We have had professional development instructors come to us about putting their courses online. We have had P.D.s and A.A.s about how do we get our students

by cohorts into Prometheus because we want to communicate to them. So, I think that they see the need or possible benefits of technology.

And students, I don't know. I think students just tend to love gadgets anyway, as long as there is someone there to teach them, as long as there is support for when the system has problems. But, I think that part of it is this generation that . . . you know, they know how to program the VCR. They know how to set up the system for video games on the television. I don't think that they are opposed to it. In fact, they have come to me when they can't remember their password, and they will say so-and-so doesn't have their course on. I don't have a four-digit password. And I say that they will get up to using it.

But again, I just think the overall thing with institutions of higher ed, faculty who are leading the institution, have an obligation not to get left behind. So that when the student gets a job, the student knows advanced PowerPoint. The student knows how to beam something with their Palm Pilot to another Palm Pilot user. I mean, they are learning it all here, even more so than any other college or university.

So, with regards to staying abreast with technology, whether it helps with learning to write a five-paragraph essay, it is probably not that useful.

CNS: The Management Institute, is that a training thing for future managers?

Faculty Member B: The senior management has decided it will take place every year, and even some years will overlap starting cohorts; they will handpick, not really, senior management and department heads, nominated people to be in the Management Institute. Some people applied, and they were not accepted.

	They are being trained to be teacher-leaders, senior management at the college.
CNS:	It sounds like a wonderful program.
Faculty Member B:	It really is a groundbreaker. But it is part of the whole corporate trend, it seems, because, boy, I have been reading a lot of books on developing leadership within your own organization. So. . . .
CNS:	That's very interesting. See, I learn something new every time I meet with you. So, tell me a little about your interests in technology. When you were at [the old building], there wasn't a lot.
Faculty Member B:	Every time technology has been introduced at this college before I took this position, I was fearful and anxious. I would hear this word *electronic,* and I was fearful. In fact, there were other faculty who just loved it. They would sit with you and help you get into it. They would be there to answer your questions.

When we were told we had to do faculty webpages, I was really anxious. And to this day, I still don't know how to create a webpage. And I think most of the faculty was on the same page as I was. We received some training from the college, but it wasn't enough to get us really rolling. So, the webpages caused a lot of us to get gray hairs, including myself. We were scared to death. Because again, the message was it is a must. Everyone must do it. And we tried. But once someone set the page up for us, we never could get back in. We didn't know what to do. Now, this position was faculty development, and I liked the idea of helping faculty, especially practitioners, to ease into teaching and to make sure that every class here, not one would be taught in such a bad way that the students would have a gap in knowledge.

Appendix M ■

The Analytical Process of Coding ■

By Carol Burg

(from her 2010 study on mentoring)

Original Transcript Meaning Units Highlighted	Theme of Meaning Units	Emergent Central Themes	Emergent Codes
Prof Intentional 06.01.09 **Carol**: (0:02) You talk. Go ahead. **Professor Intentional**: (0:02) Okay, hello, bye. **Carol**: (0:04) Good levels. **Professor Intentional**: (0:05) Okay. **Carol**: (0:07) Excellent. All right, well, today is June 1, 2009, and it is 9:07 a.m., and oh, what pseudonym would you like to use? **Professor Intentional**: (0:18) What pseudonym?			

(Continued)

(Continued)

Original Transcript Meaning Units Highlighted	Theme of Meaning Units	Emergent Central Themes	Emergent Codes
Oh, my. I don't know. I can't . . . let's see if something evolves, you know, over the course of our chat. (0:27) (laughter) **Carol**: (0:28) Okay, right now, I will call you Professor Intentional. **Professor Intentional**: (0:31) Okay, that sounds good; that sounds good. (0:33) (laughter) **Carol**: (0:33) I am here with Professor Intentional, and thank you very much for volunteering to let me interview you for my study. I am happy to report that you were one of the more highly nominated mentors by doctoral students, doctoral graduates. Intentional people nominated you, and they allowed me to also share their names with you.			

Original Transcript Meaning Units Highlighted	Theme of Meaning Units	Emergent Central Themes	Emergent Codes
(1:01) Their names are J. O., J. B. M., and J. G. So, I am very grateful that you are available to let me talk to you about mentoring. **Professor Intentional**: (1:17) Terrific. **Carol**: (1:18) Thank you. First of all, can you describe to me how you view your role as a mentor to doctoral students? **Professor Intentional**: (1:29) Well, I shared with you that chapter that I wrote on mentoring. **Carol**: (1:35) Yes, I did read it. **Professor Intentional**: (1:36) That was published not too long ago, actually, very recently. As I was doing the research for that chapter, I came across the term *intentional mentoring*. That is a term, I think, that really kind of	Prof. Intentional sees her mentoring role as an intentional one.	Mentors see their mentoring role as an intentional one.	Mentoring: Intentional

(Continued)

(Continued)

Original Transcript Meaning Units Highlighted	Theme of Meaning Units	Emergent Central Themes	Emergent Codes
embraces my philosophy on mentoring. (1:58) It is intentional. It is conscious. Now, there are many, many things that happen in a mentoring relationship that you don't necessarily think about very consciously. It is serendipity. It is something that happens in a hallway. It is something that happens in passing conversation.			
(2:15) But, I would classify my particular view of it as intentional mentoring. I try to make very sure that when I see opportunities that come through, I think, "Which of my doctoral students would benefit from this? Who could contribute to this? How can I position them for a job later on?	Prof. Intentional sees her mentoring role as an intentional one. Prof. Intentional sees part of her role as a mentor as positioning her protégés for jobs later on.	Mentors see their mentoring role as an intentional one. Mentors see part of their role as a mentor as positioning protégés for jobs later on.	Mentoring: Intentional Mentors: protégé job assistance

Original Transcript Meaning Units Highlighted	Theme of Meaning Units	Emergent Central Themes	Emergent Codes
What does this person need that this person may not need as much of?" (2:41) You know, whether it is exposure at a conference. Whether it is a writing opportunity. Whatever it is. So, it is intentional. I really try to take stock of who the doctoral students are, what their strengths are, what their needs are, and then I try to fashion the opportunities that I make available and how I interact with them a very conscious choice.

(3:07) Does that make sense? **Carol**: (3:08) Yes. Yes, it does. Thank you. Typically, what is the mentoring experience like for you? I am doing a phenomenological study, so I want to kind of get into your head and understand what your experience is, what it is like. | Prof. Intentional sees her mentoring as responding to the individual needs of protégés.

Prof. Intentional sees her mentoring role as an intentional one. Prof. Intentional sees her role as a mentor as being a conduit for opportunities for her protégés. | Mentors see their mentoring as responding to the individual needs of protégés.

Mentors see their mentoring role as an intentional one.

Mentors see their role as a mentor as being a conduit for opportunities for protégés. | Mentoring: Individual Mentors: meet needs

Mentoring: Intentional

Mentors: conduit of opportunities |

(Continued)

(Continued)

Original Transcript Meaning Units Highlighted	Theme of Meaning Units	Emergent Central Themes	Emergent Codes
Professor Intentional: (3:28) What is it like for me in terms of just my experiences of it, because the first word that pops into my mind, and if we are doing free association, Freudian analysis, is rewarding. It is very rewarding. It is a feeling like . . . well, it is a lovely thing to see people achieve, grow, and develop, and go on to do the things that they are capable of doing. So, it is very rewarding to see that. (4:00) The other thing that happens for me is that I see it as a very circuitous, kind of full-circle kind of thing. Every time, just about, that I engage in some kind of mentoring relationship or experience, I can't help but reflect back on	Prof. Intentional feels that mentoring is very rewarding. Prof. Intentional feels that it is a "lovely thing" to see protégés grow and achieve, and that is rewarding to her. Prof. Intentional sees a circuitous, full-circle flow between her mentor and her mentoring.	Mentors feel that mentoring is rewarding. Mentors feel that it is rewarding to see protégés grow and achieve. Mentors see a circuitous, full-circle flow between themselves and their mentoring.	Mentoring: Rewarding Mentors: enjoy protégé growth Mentors: enjoy protégé success Benefit: good feeling Mentors: synergistic flow

Original Transcript Meaning Units Highlighted	Theme of Meaning Units	Emergent Central Themes	Emergent Codes
the mentoring that I received, and I had a very strong, very powerful mentor whose name was Jan Lewis, Jan L. Tucker. T-U-C-K-E-R. In fact, I keep his photo here as a reminder. He passed away about 10 years ago very suddenly of a heart attack.	Prof. Intentional had a mentor.	Mentors often have a mentor.	Mentors: present/ absent
Carol: (4:43) Sorry to hear that. **Professor Intentional:** (4:44) And I think about him all the time, much more than I ever thought that I would while I was in the mentee relationship with him. He was very much the intentional mentor. He's very much the person who would say . . . he would call me into his office and say, "This conference is coming up. You need to be there. You need to go."	Prof. Intentional sees the mentoring role as an intentional one.	Mentors see the mentoring role as an intentional one.	Mentoring: Intentional

(Continued)

Original Transcript Meaning Units Highlighted	Theme of Meaning Units	Emergent Central Themes	Emergent Codes
(5:05) Sometimes, it would be a matter of time. Sometimes, it would be a matter of money, and whatever it is, he would find a way to facilitate that for me. So, he was a really great mentor. He was a leader in global education. He was from Indiana and had a very parochial background, kind of insulated. And how this man grew up to be somebody really well-known in global education is kind of an interesting trajectory. (5:36) He had a very, very high success rate with students of color, with minority students. So, when I wrote that piece that you read, almost all of the people there in that sample, almost all of them were from all kinds of walks of life, international students and what have you.	Prof. Intentional's mentor's role was to facilitate resources for her to participate in pro. dev. events.	Mentors often see their role as facilitating resources for protégés to participate in pro. dev. events.	Mentors: facilitators Mentors: protégé resource assistance

Original Transcript Meaning Units Highlighted	Theme of Meaning Units	Emergent Central Themes	Emergent Codes
(5:56) So, I see it as a kind of circuitous thing. I always go back, and I reflect on what he was able to provide for me and how he might handle a situation or what he might do or say. So, I think those are the two things. It is like a pay it forward but also a reflection back.	Prof. Intentional sees a circuitous, full-circle flow between her mentor and her mentoring. Prof. Intentional "pays forward" her good mentoring and reflects back on her good mentoring.	Mentors see a circuitous, full-circle flow between themselves and their mentoring. Mentors "pay forward" their good mentoring. Mentors reflect back on their good mentoring.	Mentors: synergistic flow Mentors: pay forward their good mentoring Mentors: reflect on their mentor
Carol: (6:15) Interesting. Interesting. So, if I understand what you are saying, it seems that in your daily experience, mentoring seems to be present in your daily thoughts, like when you're going through your academic day and you come across articles or opportunities. It is almost as if one of the first things that pops into your head is, "Oh, this might help so-and-so, or this would be good for so-and-so."			

(Continued)

Original Transcript Meaning Units Highlighted	Theme of Meaning Units	Emergent Central Themes	Emergent Codes
Professor Intentional: (6:54) Daily. That is a daily occurrence. That is exactly right. It can be an article when I am doing my own lit review for a piece that I am working on. I'll come across something. And it is nothing to take an extra minute maybe to download an article, save it on your desktop, open up an e-mail, and then forward it and attach it to send it to somebody. That is a very classic thing. (7:19) Other times, it might be a colleague, a former colleague of mine	Prof. Intentional sees her mentoring as a daily occurrence and easily integrated into her daily routine. When Prof. Intentional finds an article that might help a protégé, she takes a minute to forward it to the protégé.	Mentors see their mentoring as a daily occurrence and easily integrated into their daily routines. When mentors find an article that might help a protégé, they takes a minute to forward it to the protégé.	Mentoring: Easily Accommo-dated Mentoring: Integrated Mentors: frequent mentoring Mentoring: Frequent Mentoring: Intentional Mentors: helpers
coincidentally, the lady in the picture. She is now at another U. So, she is the editor; she is going to be guest editor for *Social Education* in our field. She is editing a special issue on human rights	Prof. Intentional sees her role as a mentor as being a conduit for opportunities for her protégés.	Mentors see their role as a mentor as being a conduit for opportunities for protégés.	Mentors: conduit of opportunities

Original Transcript Meaning Units Highlighted	Theme of Meaning Units	Emergent Central Themes	Emergent Codes
violations around the world. So, she sent me an e-mail saying would I contribute an article to the journal? And I said, "Well, maybe I could, but any problem with my forwarding this call for papers to my doctoral students?" (7:55) She goes, "Oh, no, not at all. Great." So I did. I forwarded it to all of our doctoral students, and then several of them submitted— developed a prospectus for her. In fact, just yesterday . . . so I got a few of them onboard on that. Then, one of the doctoral students last night said, "Here's the manuscript. Before I send it to Dr. K., any chance that you could take a look at this and give me feedback and edit or whatever?" (8:27) So, of course. Of course. There is no question. So, that would be another example.	Prof. Intentional gives feedback & edits her protégés' manuscripts for publication.	Mentors give feedback & edit protégés' manuscripts for publication.	Mentors: assist protégé publication

(Continued)

(Continued)

Original Transcript Meaning Units Highlighted	Theme of Meaning Units	Emergent Central Themes	Emergent Codes
In other cases, it is a doctoral student proposed . . . defended her proposal on Friday successfully. So, after she finished, I said, "I'd like for you to come back to my office so that we can come up with a timeline of what the next year might look like so that we can position you to find jobs." (8:56) So she did. We broke it down month by month. So then this morning, I went to my colleague, who is always searching the "Chronicle" and stuff. I said, "Look, I just want to let you know, Carrie defended on Friday. Any jobs that you come up with, she is really looking at Kentucky, Tennessee, the Carolinas, that kind of corridor; can you send them?"	Prof. Intentional sees part of her role as a mentor as positioning her protégés for jobs.	Mentors see part of their role as a mentor as positioning protégés for jobs.	Mentors: protégé job assistance

Original Transcript Meaning Units Highlighted	Theme of Meaning Units	Emergent Central Themes	Emergent Codes
(9:18) So, it is a daily thing. It is a daily thing. It is a daily thing. And it really doesn't take a whole lot of time because I think that is what sometimes makes faculty members shy away from this. They think that it is a very time-consuming thing. It is part of what you normally do. But, it has to be conscious. You have to be aware that this . . . you have to internalize this as one of your jobs, if you will, one of your duties. (9:44) So, when I write my annual review, my annual report every year, there is a . . . I get a very small—it's too small if you ask me; I think the faculty should have a larger assignment in this regard—but I do have what is	Prof. Intentional sees her mentoring as a daily occurrence and easily integrated into her daily routine. Prof. Intentional sees her mentoring as a daily occurrence and easily integrated into her daily routine. Prof. Intentional sees the mentoring role as an intentional one.	Mentors see their mentoring as a daily occurrence and easily integrated into their daily routines. Mentors see their mentoring as a daily occurrence and easily integrated into their daily routines. Mentors see the mentoring role as an intentional one.	Mentoring: Easily Accommo- dated Mentoring: Integrated Mentors: frequent mentoring Mentoring: Frequent Mentoring: Easily Accommo- dated Mentoring: Integrated Mentors: frequent mentoring Mentoring: Intentional

(Continued)

(Continued)

Original Transcript Meaning Units Highlighted	Theme of Meaning Units	Emergent Central Themes	Emergent Codes
called the other instructional effort. So, it's not what you teach; it's other instructional effort. (10:06) But in my mind, I translate that as doctoral student mentoring. I don't see that as service to my profession. I don't see that as research. That is not teaching, but is other instructional effort. Since I have that kind of consciousness, that that's part of what I do, as part of my job, it happens. It takes place.	Prof. Intentional sees doctoral student mentoring as "other instructional effort"—as part of her job.	Mentors see mentoring as part of their jobs.	Mentoring: Professional Responsibility/ Part of Job

Bibliography ■

Admiraal, M. (2000). *Overcoming the double challenge: Reflections on Phil's experiences.* Unpublished critical case study, Eastern Michigan University at Ypsilanti.

Beltman, T. (2002). *Missing dad.* Unpublished critical case study, Eastern Michigan University at Ypsilanti.

Berg, B. (2007). *Qualitative research methods for the social sciences* (6th ed.). Boston: Allyn & Bacon.

Best, J. (2004). *More damned lies and statistics: How numbers confuse public issues.* Berkeley: University of California Press.

Bradley, C. (2009). *Reflection on the Bean Depot observation.* Unpublished class assignment, University of South Florida at Tampa.

Burg, C. A. (2010). *Faculty perspectives on doctoral mentoring: The mentor's odyssey.* Unpublished doctoral dissertation, University of South Florida–Tampa.

Chuan, C. T. Y. H. (1963). *The mustard seed garden manual of painting.* Princeton, NJ: Princeton University Press.

Csikszentmihalyi, M. (1996). *Creativity: Flow and the psychology of discovery and invention.* New York: Harper.

Dewey, J. (1938). *Experience and education.* New York: Collier.

Dewey, J. (1958). *Art as experience.* New York: Capricorn.

Dewey, J. (1967). *The early works.* Carbondale: Southern Illinois University Press.

Edwards, B. (1979). *Drawing on the right side of the brain.* Los Angeles: J. P. Tarcher.

Edwards, B. (1986). *Drawing on the artist within: A guide to innovation, invention, imagination and creativity.* New York: Simon & Schuster.

Edwards, B. (1999). *The new drawing on the right side of the brain.* Los Angeles: J. P. Tarcher.

Einstein, A., & Infeld, L. (1938). *The evolution of physics.* New York: Simon & Schuster.

Ellis, E. R. (1995). *A diary of the century: Tales from America's greatest diarist.* New York: Kodansha International.

Estes, C. P. (1992/1996). *Women who run with the wolves: Myths and stories of the wild woman archetype.* New York: Ballantine.

Goldberg, N. (2005). *Writing down the bones.* Boston: Shambala Press.

Gouwens, J. (1995). *Leadership for school change: An interview observation study of leadership of two Chicago elementary principals.* Unpublished doctoral dissertation, University of Kansas–Lawrence.

Hemingway, E. (1995). *The snows of Kilimanjaro.* New York: Scribner.

hooks, b. (1994). *Teaching to transgress: Education as the practice of freedom.* New York: Routledge.

Janesick, V. J. (1991). Ethnographic inquiry: Understanding culture and experience. In E. Short (Ed.), *Forms of curriculum inquiry* (pp.101–119). Albany: SUNY Press.

Janesick, V. J. (1995). A journal about journal writing as a qualitative research technique: History, issues, and reflections. *Qualitative Inquiry, 5*(4), 505–524.

Janesick, V. J. (1998). The dance of qualitative research design: Metaphor, methodolatry, and meaning. In N. Denzin & Y. Lincoln (Eds.), *Strategies of qualitative inquiry* (pp. 35–55). Thousand Oaks, CA: Sage.

Janesick, V. J. (1999). Using a journal as reflection in action in the classroom. In D. Weil (Ed.), *Perspectives in critical thinking: Theory and practice in education.* New York: Peter Lang.

Janesick, V. J. (2000). The choreography of qualitative research design: Minuets, improvisations and crystallization. In N. K. Denzin & Y. S. Lincoln (Eds.), *Handbook of qualitative research* (2nd ed., pp. 379–399). Thousand Oaks, CA: Sage.

Janesick, V. J. (2001). Intuition and creativity: A pas de deux for the qualitative researcher. *Qualitative Inquiry, 7*(5), 531–540.

Janesick, V. J. (2007). Oral history as a social justice project: Issues for the qualitative researcher. *The Qualitative Report, 12*(1), 111–121. Retrieved December 1, 2009, from http://www.nova.edu/ssss/QR/QR12-1/janesick.pdf

Janesick, V. J. (2008). Art and experience: Lessons learned from Dewey and Hawkins. In J. G. Knowles & A. L. Cole (Eds.). *Handbook of arts in qualitative inquiry: Perspectives, methodologies, examples and issues* (pp. 477–483). Thousand Oaks, CA: Sage.

Kulavuz-Onal, D. (2008). *Reflections on observations.* Unpublished class assignment, University of South Florida at Tampa.

Kvale, S. (1996). *Interviews: An introduction to qualitative research interviewing.* Thousand Oaks, CA: Sage.

Kvale, S., & Brinkmann, S. (2009). *Interviews: Learning the craft of qualitative research interviewing* (2nd ed.). Thousand Oaks, CA: Sage.

Lincoln, Y. S., & Guba, E. G. (1985). *Naturalistic inquiry.* Beverly Hills, CA: Sage.

Lortie, D. (2002). *Schoolteacher: A sociological study* (2nd ed.). Chicago: University of Chicago Press.

Merriam, S. B. (1997/2001). *Qualitative research and case study applications in education* (Rev. ed.). San Francisco: Jossey-Bass.

Merriam. S. (2009). *Qualitative research: A guide to design and implementation* (Rev. ed.). San Francisco: Jossey-Bass.

Mishler, E. G. (1986). *Research interviewing: Context and narrative.* Cambridge, MA: Harvard University Press.

Morrison, G. (2008). *Early childhood education today* (11th ed.). New York: Pearson.

Mooney, B., & Holt, D. (1996). *The storyteller's guide.* Little Rock, AR: August House.

Morgan, D. (Ed.). (1993). *Successful focus groups.* Newbury Park, CA: Sage.

Pepe, J. (2007). *Reflection: Skills, difficulties, and what I would change.* Unpublished class assignment, University of South Florida at Tampa.

Piantanida, M., & Garman, N. B. (1999). *The qualitative dissertation: A guide for students and faculty.* Thousand Oaks, CA: Corwin.

Progoff, I. (1992). *At a journal workshop: Writing to access the power of the unconscious and evoke creative ability.* Los Angeles: J. P. Tarcher.

Rainer, T. (1978). *The new diary.* New York: G. P. Putnam.

Roulston, K. (2010). *Reflective interviewing: A guide to theory and practice.* Thousand Oaks, CA: Sage.

Rubin, H. J., & Rubin, I. S. (2005). *Qualitative interviewing: The art of hearing data* (2nd ed.). Thousand Oaks, CA: Sage.

Saldana, J. (2009). *The coding manual for qualitative researchers.* Los Angeles: Sage.

Schwandt, T. A. (2001). *Qualitative inquiry: A dictionary of terms* (2nd ed.). Thousand Oaks, CA: Sage.

Slotnick, R. (2010). *University and community college administrators' perspectives of the transfer process for underrepresented students: Analysis of policy and practice.* Unpublished doctoral dissertation, University of South Florida–Tampa.

Spradley, J. P. (1980). *Participant observation.* New York: Holt, Rinehart & Winston.

Stake. (1995). *The art of the case study.* Thousand Oaks, CA: Sage.

Stevenson, C. (2002). *A case study on educational leaders' perspectives on technology use in the undergraduate general education curriculum.* Unpublished doctoral dissertation, Roosevelt University–Chicago.

Szent-Gyorgi, A. (1996). *The Albert Szent-Gyorgi papers.* Bethesda, MD: U.S. National Library of Medicine.

Tharp, T. (2003). *The creative habit: Learn it and use it for life.* New York: Simon & Schuster.

Toloday, K. (2002). *Follow the leader.* Unpublished critical case study, Eastern Michigan University at Ypsilanti.

Twain, M. (1996). *The wit and wisdom of Mark Twain.* Philadelphia: Running Press Book.

Tye, J. (2002). *Being an adolescent is tough enough: What if you have special needs?* Unpublished critical case study, Eastern Michigan University at Ypsilanti.

Vasquez, A. (2009). *Interview protocols.* Unpublished class assignment, University of South Florida at Tampa.

Vrobel, O. (2009). *Sample reflection for adding to the reflective journal: Self-evaluation upon observing in a student laboratory for English language learning.* Unpublished class assignment, University of South Florida at Tampa.

Williams-Boyd, P. (2005). *The critical case use of observation of a student: Lessons learned for the professional development of the educator in training.* Unpublished class assignment, Eastern Michigan University at Ypsilanti.

Wolcott, H. F. (1995). *The art of fieldwork.* Walnut Creek, CA: AltaMira.

Index ■

CPSIA information can be obtained
at www.ICGtesting.com
Printed in the USA
FFOW04n1958140116
20412FF